THE BENEFITS OF BEHAVIORAL RESEARCH TO THE FIRE SERVICE

THE BENEFITS OF BEHAVIORAL RESEARCH TO THE FIRE SERVICE

◆

HUMAN BEHAVIOR IN FIRES AND EMERGENCIES

Peter W. Blaich

iUniverse, Inc.
New York Lincoln Shanghai

THE BENEFITS OF BEHAVIORAL RESEARCH TO THE FIRE SERVICE
HUMAN BEHAVIOR IN FIRES AND EMERGENCIES

iUniverse books may be ordered through booksellers or by contacting:

iUniverse
2021 Pine Lake Road, Suite 100
Lincoln, NE 68512
www.iuniverse.com
1-800-Authors (1-800-288-4677)

Because of the dynamic nature of the Internet, any Web addresses or links contained in this book may have changed since publication and may no longer be valid.

ISBN: 978-0-595-48549-9 (pbk)
ISBN: 978-0-595-60646-7 (cloth)
ISBN: 978-0-595-60643-6 (ebk)

Printed in the United States of America

This book is dedicated to my wife Irene Mary Blaich and my sons William Peter Blaich and Adam Peter Blaich. Without my wife I would have never made it through the days, weeks and months after the terrorist attacks of September 11, 2001; she did and continues to give me strength and support in this very demanding career. My sons Billy and Adam I look in their eyes and I see hope and promise in a very troubled world.

Contents

Preface xi

About This Book. xiii

Acknowledgements xv

Fire Department of New York (FDNY) Joint Program with John
    Jay College xvii

Foreword. xix

CHAPTER 1   Human Behavior As It Pertains To Fire Incidents 1

    Introduction 1

    Fire Related Factors That Differ Among the United States and Other Modern
        Industrialized Nations 2

    Identifying Populations, Structures, and Communities at High Risk 5

    Progressive Movement of the Study of "Fire Related Human Behavior" in all Its
        Applicable Aspects. 9

    A New Change in Direction Driven By Research 17

    Chapter One Summary. 19

    Discussion and Review 20

CHAPTER 2   Trusting Research and Putting Faith in What It
                 Tells Us. 22

    Introduction 22

    Techniques That Appeal To Emotion Instead Of Reason 23

    The Various Techniques Used To Study Human Behavior. 27

    Human Behavior Research Design 29

    Analyzing the Relationship between Variables. 42

    Cost-Benefit Analysis 46

    Chapter Two Summary. 49

Discussion and Review... 56

CHAPTER 3   A Systems Approach To Fire Related Human
Behavior....................................... 59

Introduction 59

A Systems Approach That Incorporates Human Behavior 61

A Fresh Look at Fire Protection Engineering 63

Goal-Based Approaches 71

A Comprehensive Look at the World Trade Center Evacuation............... 76

An Evacuation Model for High-Rise Buildings 84

Private Dwelling Fires 89

The Uniqueness and Compromise of Residential Care Facility 92

Summary... 98

Discussion and Review... 99

CHAPTER 4   Fire Safety Design and Fire Investigation 101

Introduction 101

Fire Safety Design and Local Regulations.............................. 101

Group Response to Fire.. 103

Firefighter Fatalities Reports 108

Fire Analysis for Fire Investigation.................................... 111

Behavioral Factors of Arson 114

Summary... 117

Discussion and Review... 117

APPENDIX A   Short Staffing: A Congressional Priority......... 119

APPENDIX B   Risk Perception and Security Bars 123

APPENDIX C   A Strong Preference for using Familiar Routes to
Escape and Reentry behavior 125

Endnotes 127

About the Author 131

Preface

The study of "Human Behavior in Fires and Emergencies" is multifaceted. Of significance, having an understanding of "Human Behavior in Fires and Emergencies" is the most valued ability to be gained through a collegiate Fire Science Program. Regardless of how many components of the fire protection delivery system you have studied, you will find that the study of "Human Behavior in Fires and Emergencies" provides invaluable information on the behavioral implications of fire incidents. Moreover, the study of human behavior and fire emergencies is also significant because, as fire safety professionals, you want to know what to expect from people that find themselves in a fire situation. Ultimately, understanding the dynamics of human behavior will help you do your job more effectively as someone either employed in public fire protection or operating as a private consultant working in the field of fire protection.

About This Book

By reading this book the student will learn about the impact of fire in the United States as compared to other industrialized nations. Additionally, this book will explore the social and economic factors that influence fire rates. Moreover, since the study of human behavior relies heavily on research methods, the book will help you sharpen your research skills as well as your skills for conducting and evaluating existing research. Of significance, this book will introduce you to a systems approach that many Certified Fire Protection Specialists and Fire Protection Engineers use to help them understand the interactions of humans that untimely find themselves in fire situations.

Acknowledgements

I would like to personally thank my Grandfather Charles Blaich, my Father William Blaich and my Uncle Charles Blaich all retired New York City Chief Fire Officers. They are the people that inspired and showed me what a rewarding career firefighting can be with the Fire Department of New York. Moreover, they continue to encourage me to be the best that I can be. Additionally, I thank Chief Fire Officer Ronald Spadafora (FDNY) and Professors Glen Corbett, Norman Groner and Ned Benton at John Jay College's Division of Fire Science for mentoring me and helping me to develop myself as a professional. Ultimately, they have increased my knowledge of Fire Science as well as Fire Administration and Fire Protection Management. In doing this the aforementioned people have elevated the professional standing of all firefighters in the eyes of the citizens of New York City.

Fire Department of New York (FDNY) Joint Program with John Jay College

FDNY and John Jay College of Criminal Justice has combined resources to present a series of college courses that can be applied toward a Fire Science Bachelor of Science Degree or a Fire and Emergency Service Bachelor of Arts Degree. Each course carries three credits, and is taught by FDNY personnel and is offered at the Bureau of Operations Training Facility at Randall's Island and Fort Totten. Classes meet for 15 weeks, one night a week.

Today's managers need expertise in the fields of fire protection, emergency planning and security to protect the assets of their organizations or communities. John Jay College of Criminal Justice has answered this demand with a unique master's degree program in protection management. John Jay College is the only graduate school where you take courses in three of the disciplines charged with protecting the lives and property of organizations and communities. Additionally, you take key courses that provide you with the knowledge and skills to be an effective manager. The Master of Science in Protection Management Program combines theory with practice. John Jay supply students not only with the current state of knowledge, but also with the tools and theoretical background to stay up to date. The faculty is unusually capable most of the courses are taught by full-time faculty who are among the most important leaders and thinkers in their specialization either fire protection, emergency planning or security. Currently the fire science division at John Jay College shares its management courses with the Department of public management so that students learn how to be an effective manager from a faculty that is arguably the best in the college that specialize in criminal Justice or fire science. Students study the three fields of responsibility for protecting the assets of organizations and communities, while concentrating in Fire Protection, Emergency Planning or Security. Of significance, students taking management classes from John Jay College's Department of Public Management rated #1 by U.S. News and World Report.

For information about John Jay College's Batchelor Degrees in Fire science or Fire and Emergency Services please contact the program director Professor Glen Corbett at phone number 212-237-8092 or drop an e-mail to him at gcorbett@jjay.cuny.edu.

For information about John Jay College's Master of Science Degree in Protection Management please contact the program director Professor Norman Groner at phone number 212-237-8834 or drop an e-mail to him at ngroner@jjay.cuny.edu.

Or visit the official college website at: http://web.jjay.cuny.edu.

John Jay College is located in midtown Manhattan at corner of 59th Street and 10th Avenue. John Jay College is accredited by the Higher Learning Commission as well as the Middle States College Association.

Foreword

In the aftermath of the terrorist attacks and subsequent conflagration of September 11, 2001 Fire Protection Specialist across the board are drawing on Behavioral Science to better understand people's reactions during fire and emergency situations; in an effort they hope will lead to safer buildings and more manageable evacuations. Previously, fire safety engineers worked under a simple assumption that when a fire alarm activated, people will evacuation immediately. It was believed that how quickly people managed to evacuate a building depended mainly on physical abilities as well as the location of the nearest exit and the behavior of the fire. But work by behavioral scientists has found that this idea falls considerably short. Research now shows that as much as two thirds of the time it takes occupants to exit a building after the fire alarm sounds is startup time or time spent milling about and looking for more information. Ultimately, such a finding has big implications for architects, engineers and fire protection specialists hoping to design safer buildings. Subsequently, after the 9/11 terrorist attacks, this new way of thinking based on human behavior in fires and emergencies is getting more attention and funding. In fact, a number of behavioral scientists have landed government grants from both the US Centers for Disease Control and Prevention and the National Institute of Standards and Technology to learn about human behavior in large-scale building evacuations by studying what happened at the World Trade Center during the attacks and subsequent fires. Studying how occupants reacted as events unfolded and finding out what helped or hindered the evacuation efforts could provide invaluable information for future building designs. Ultimately, it is the author of this book's belief that this is going to impact structured emergency preparedness in a major way.

Psychologists have been studying how people react during fires for more than 25 years. The late 1970s and early 1980s were a particularly productive time, says Norman Groner, Ph.D., a Psychologist who studied Human Behavior in Fires since the 1980s. At that time, he says, the National Bureau of Standards which was the precursor to today's National Institute of Standards and Technologies funded the research through its Center for Fire Research. But much of that funding dried up in the mid-1980s, as NIST began to focus on other things like computer modeling of fire dynamics instead. The post-9/11 renewal of interest in this area is important for the work for helping to design better evacuation systems Professor Groner John Jay College of Criminal Justice notes. In fact, over the years, the research has debunked many of the myths that used to be and those that still exist.

The basic premise of this book is that design should be human-centered because human-centered design provides for the information that people need to adapt to the chaotic and uncertain way that fire develops. Ultimately, some of the recommendations resulting from the Study of "Human Behavior in Fires and Emergencies" are starting to make its way into legislation. Most recently the New York City Fire Department has completed a final draft of a new fire code, this comprehensive revision of the Cities almost a century old fire code is based on a model code, the International Fire Code, published by the International Codes Council, Inc. By enacting a fire code based on the International Fire Code, the City of New York will join a growing number of states and municipalities that across the Country has adopted a uniformed fire safety standard which address specific concerns related to human behavior in fires and emergencies.

1

Human Behavior As It Pertains To Fire Incidents

o o o o o o o o o o o o o o o o o o o o o o o o o o o o o o o o
"Behavioral research has shown that it is erroneous to assume that behavior and fires is simply a process largely controlled by exit and alarm systems."

—*Pauls and Jones in*
Fire Journal

Introduction

Professor John L. Bryan, an American researcher and Professor of Fire Engineering at the University of Maryland, is considered the founding father of "Fire Related Human Behavior." Professor Bryan conducted extensive studies on human behavior in fires involving disastrous losses in the United States. Subsequently, the Field of "Fire Related Human Behavior" received increased attention following the publication in 1973 of *America Burning*, a report by the National Commission on Fire Prevention and Control. Currently, the America Government's National Institute of Standards and Technology (NIST) an agency of the U.S. Department of Commerce is conducting large scale research on the behavior of humans and fire emergencies. Moreover, studies are being conducted by laboratories funded by universities and private sector researchers regarding the human factors involved in fire incidents. Additionally, the National Fire Protection Association (NFPA) also is active in conducting research that focuses on how people behave in fire incidents. Presently, the NFPA dispatches field teams to collect data on major fires and publishes case studies based on those investigations. International governments have also contributed to the current understanding of "Fire

1

Related Human Behavior." For example, Japan recognizes fire as a major life safety threat, and researchers in the Japanese government and in Japanese Universities are actively involved in conducting studies of "Fire Related Human Behavior."

Fire Related Factors That Differ Among the United States and Other Modern Industrialized Nations

In addition, to the Japanese, the British have devoted studies exclusively to individual and group behavior in fires. British researchers have covered a range of topics, including fire setting, fire and institutions such as hospitals and nursing homes, and problems related to evacuation. A convention featuring lectures was titled "Human Behavior in Fire" and was held in Belfast, Northern Ireland, in August of 1998. Presenters covered a wide range of topics, from computerized models that predict human movement during fire evacuations to discussions on how people make decisions faced with the ambiguous information that is common to a fire situation. An estimated 140 experts and interested persons attended and the size of the turnout surprised the meetings organizers. The excellent turnout is attributed to an interest in performance-based codes. Subsequently, several countries including the United States and Canada are rapidly developing and adapting performance-based codes. Performance-based codes allow architects and fire protection engineers to plan buildings based on models and calculations that show how well their designs are expected to protect occupants during various types of fire incidents. Those developing such codes expect them to allow regulatory authorities to approve otherwise safe buildings in instances where following the exact provisions of traditional perspective codes would prohibit that construction. Because using traditional perspective codes, building architects do not need to pay much attention to human behavior. Alternatively, it is difficult to justify a performance-based design without carefully attending to the behaviors of the buildings occupants. Ultimately, to realize fully the promise of performance-based fire safety designs, the field needs more as well as better research and theory on how humans behave in fire situations.

The United States of America has a never ending disastrous fire problem, even more than generally perceived. Yearly, America is stricken with millions of fires resulting in thousands of deaths, tens of thousands of injuries, and billions of dollars of loss subsequently making the U.S. fire problem an ongoing epidemic. Of significance, the indirect cost of fire increases the magnitude of economic losses tenfold. Ultimately, the United States continues to have fire death rates and property losses that are among the largest of the industrialized modern nations. According to the United States Fire Administration (USFA) the Nation had a yearly average of 1,872,800 fires and 4266 fire deaths from the year 1992 to 2001. There were 6,196 fire fatalities in the year 2001 alone and of significance this statistic includes the tragic events of September 11, 2001. Unfortunately, from a statistical point of view the injury statistics are not as clear-cut as the death tolls because of ambiguity about the completeness of the finding and reporting minor injuries in the fact that many injured people go directly to the medical facility themselves without being reported to or treated by the fire department or local emergency medical service responders which maintain the numbers. Civilian injuries from reported fires average 24,900 per year over a 10 year period. Firefighter injuries averaged 46,200 from those fires, and a study suggests that the number of civilian injuries associated with these fires that are not reported to the fire service might be several times that of the number from reported fires. In terms of financial losses, the estimated direct value of property destroyed in fire was 44 billion including the World Trade Center losses. These casualties in losses come from an average of nearly 2 million fires annually. Of significance, reported fires increased slightly in the year 2001 according to the United States Fire Administration.[1]

The fire problem is much larger than generally known. Deaths and injuries from all natural disasters combined floods, hurricanes, tornadoes; earthquakes, etc. are just a fraction of the annual casualties from fire. Deaths from disasters are on the order of 200 to 250 per year versus more than 4,000 deaths from fires. Cumulatively, fires are relatively small, and there impact is not easily recognized. Only a few fires each year have a huge dollar loss that is associated with tornadoes, hurricanes, or floods. The Southern California wild land fire is in the fall of 1993 resulted in over 800 million in losses. The Oakland East Bay Hills fire of October 1991 was estimated to have cost 1.5 billion in losses. The 1998 Florida wild land fires resulted in 390 million in timber losses. But because of the losses from fire are spread over the nearly 2 million fires that are reported each year, the total loss is far more than the impression many people have of it from the anecdotal reporting of local fires in the news media. According to the National Safety Council fires also are an important cause of accidental deaths. Fires are the fifth leading major course of accidental deaths, behind vehicle accidents, falls, poisoning, and accidental threats to breathing, which include suffocation, accidental ingestion or inhalation of objects that obstruct the airway and the like accidental drowning are not included. Fire related injuries to civilians and firefighters number over 100,000 and possibly two or three times that many when injuries from unreported fires and unreported injuries from reported fires are taken into account. Burn injuries are particularly tragic because of the tremendous pain-and-suffering that occurs. Serious burns tend to cause psychological damage as well as physical damage and they may well involve not only the victims but also their families, friends and fellow workers.[2]

Even though much progress has been made and the death rate is less than half what it was in the late 1970s and down 30 percent since 1992, the United States has a fire death rate two and half (2 ½) times that of several European nations and at least twenty (20%) percent higher than many. The U.S. fire death rate, averaged for the years 1998 to 2000, was reported at 15.5 deaths per million populations. Switzerland rated the lowest of the European nations with 6.4 per million populations, Sweden's was 15.3. Of the 25 industrial modern nations examined by the World Fire Statistics Center, the U.S. rate is still in the upper tier 20th of the 25. Moreover, this general status has been unchanged for the past 20 years. One theory of explanation regarding the high U.S. fire death rate is that the United States has placed greater emphasis on fire suppression than any other nation, whereas other nations tend to surpass the United States in practicing fire prevention. The United States would be well served by studying and implementing international fire prevention programs that have proven effective in reducing the number of fires deaths in other countries. The United States has excellent building technology; public buildings generally have good records. It is the combination of safety built into homes and safe behavior in homes in particular private dwellings that we fall short of some nations. We have the technology and installed sprinkle systems and knowledge of compartmentation but they are not used within private dwellings.[3]

The total cost of fire to society is overwhelming and estimated at over 165 billion per year. This includes the course of adding fire protection to buildings, the cost of paid five departments, the equivalent cost of volunteer departments which is annually 20 billion; the cost of insurance overhead as well as the direct cost of fire related losses and medical costs of fire injuries and other direct and indirect costs. Even if these numbers are high by as much as 100 percent, the total cost of fire ranges from 50 to 100 billion, still enormous, and on the order of one to one half percent of the gross domestic product, which was 10.1 trillion in 2001. As we can see from this economic viewpoint, fire ranks as a significant national epidemic.[4]

America is lagging behind, and continues to burn despite all the best efforts; and despite the wealth of advanced fire suppression technology as well as the best fire service delivery mechanism in the world. Some of the reasons that contribute to a higher death rate in the United States are for starters the international community focuses on prevention and not mitigation. Most other countries have far more staff involved in prevention. Most countries have a national standard for fire prevention codes. Awareness regarding the destructive forces of fire is prevalent in most countries outside the United States. The perception of fire in the United States is that fire is a usual occurrence and it is perceived as a cultural norm. Firefighters in countries outside of the United States received more training and in particular training in fire prevention. Of significance, fire service personnel afford a higher professional status in most countries throughout Europe. Fire prevention education receives a greater budgetary expenditure throughout Europe. Countries outside the United States exercise more effort targeting specific populations through fire prevention programs and campaigns. All-in-all this is just the tip of the iceberg as to why the United States is lagging behind in fire safety.

Identifying Populations, Structures, and Communities at High Risk

Unfortunately the fire problem is more severe for some groups than others. Higher risk populations include males, the elderly, African Americans, American Indians, and the very young. Even as fire casualties in the United States continue in a downward trend these groups remain at high risk. According to the United States Department of Commerce the highest states in 2001 for fire casualties were Arkansas, Delaware, Mississippi, and Alaska. The lowest were Nevada, California, Hawaii, and Utah. Not surprisingly, large population states are at the top of the list. A notable exception in 2001 is California, traditionally one of the 10 states with a higher number of deaths. In 1998, California was the third-highest state with 191 fire fatalities; in 2001, only 78 deaths were reported by the state fire marshal. As in previous years, the 10 states with the most fire deaths accounted for nearly half of the national total. Unless they're fire problems are significantly reduced, the national total will be difficult to lower. The Southeastern United States continues to have the highest fire death rate in the nation and one of the highest in the world. Although the fire death rates of the Southeastern states continue to decrease along with the overall U.S. rate, most still have death rates of 17 or more deaths per million populations, with the notable exception of Florida. In addition to the two Southeast states of Arkansas and Mississippi, Alaska and Delaware were in the highest fire death rate category in 2001. The Southeast and Alaska have been among the highest fire death rate areas for many years. In contrast states with less than 11 deaths per million population, in the same range as the nations of Europe and the Far East, tend to be states in the South West and West, like Florida, New York, New Jersey, Massachusetts, Minnesota, and Wisconsin which all had low rates in 2001. Of significance, California and Florida continue to have the lowest death rates among the high population states as they have had for many years.[5]

Not as many women die in fires than men. Research shows that a high proportion of male to female fire deaths has been remarkably steady. Males continue to have a higher fire death and injury rate per million populations than females for all age groups. Males age 15 to 50 have twice the fire death rate as women in 2001. Males in general have fire death rates one half to two times that of females. It is not known for certain the reason for the disparity of fire injuries between men and women. Stereotypes include the greater likelihood of men being intoxicated, the more dangerous occupations of men, most industrial fire fatalities are male and most firefighters are male, and a frequent handling of gasoline and other viable liquids by men. Moreover, we also know that men have more injuries trying to extinguish the fire and rescue people than do women.

The elderly in particular people over 65 years of age have a much higher fire death rate than the average population. Children under age 5 have a much greater risk of death than other children; children over five have less than average risk. In 2001, the risk of fire deaths dropped off sharply between five and 14, then slowly increased until age 64. At age 65, the risk of death by fire rose substantially above average and continued increasing as the population aged. It can be said that these profiles have remained relatively constant through the years. In contrast the age profile for injuries is very different from that of fire deaths. The risk of injury from fire is highest for adults aged 20 to 44 and the elderly over 85. The risk of injury is below average for children age 5 to 14 and for those ages 50 to 79. Those under age 5 accounts for 12 percent of the deaths with age reported, the highest proportion for any age group. This unfortunately represents a three percentage point increase from 1998 for this young age group. Those 65 and over comprise 23 percent of the fire deaths. These two peak risk group represent over one third of all fire deaths. On the other hand, two thirds of fire deaths fall in age groups that are not at high risk. Simply put the bulk of fire deaths occur to the not so young and not so old. Ultimately, the programs aimed only at the highest risk groups will not reach the majority of potential victims. Unlike the age distribution of deaths, the injury age distribution tracks closely to the relative risk profile by age. The exception to this, however, is the elderly. Ages 20 to 44 which account for nearly half of the 2001 fire injuries. The young, under age 10, account for 10 percent; the elderly over age 70 account for 9 percent. Although the elderly are at high risk, there are a few of them in total population. If their risk continues to be the same, we could expect more and more elderly fire victims and deaths as the elderly proportion of the population increases. Meanwhile, the focus for injury prevention should be on adults aged 20 to 44. It is believed that males in this age group are greater risk takers during fires, resulting in a higher proportion of injuries. The distribution of fire deaths and injuries by each is somewhat different for males versus females. Males tend to have a slightly higher proportion of deaths and injuries until age 60. After 60, the male to female proportions are reversed. The proportion of male fire deaths is higher in the midlife years, ages 15 to 60. By contrast, elderly females have twice the proportion of both deaths and injuries to fire than elderly males.[6]

Fire does not discriminate, the fire problem cuts across all groups and races, rich and poor, urban and rural. But notably, it is higher for some groups than for others. Research on race or ethnic group of victims are somewhat ambiguous in a society where many people are of a mixed heritage. In addition, many citizens including firefighters find it politically incorrect to report on race. Ultimately though, there does seem to be a higher fire problem for some ethnic groups, and it can be helpful to identify their problems for use within their own communities and by the fire prevention officers. African Americans and American Indians have higher fire death rates per capita than the national average. Fire victims of African American heritage comprise a large and disproportionate share of the total fire deaths. Although African-Americans comprise 12.7% of the population, they account for 25 percent of the fire deaths. In contrast, Asians have a very low death rate, male fatalities are less than 40 percent in women and fatalities are less than 20 percent of the overall average for their respective genders. As previously noted, male fire death rates exceed that of females by 1.5 to 2 times, and the elderly of all ethnic groups have the highest fire death rates. Resulting in an alarming statistic of elderly African Americans, 85 years of age and older, having the highest fire death rate in the nation at more than 14 times the U.S. average fire death rate. Ultimately, this situation is getting worse and not better. In 2001, elderly African American males had by far the highest relative risk at 21.5 times the U.S. average.[7]

Over the years, there has been little change in the proportion of fires, deaths, injuries, and dollar loss by type of property involved. Most fires occur outside to be precise 41% occur in fields, vacant lots, trash, and the forest. Many of these fires are arson, intentionally set but ultimately do not cause much damage. Fires occurring in residential and nonresidential structures together comprise only one third of the total fires, with residential fires outnumbering nonresidential structural fires by three to one. What is surprising is the large number of vehicle fires. In fact, one out of every five fires to which a fire department responds involves a vehicle. Of significance, the largest percentage of deaths, 77 percent in 2001, occurs in residence, with the majority of these fires striking one in two family dwellings. Automobiles account for the second-largest percentage of fire deaths at 16 percent. Ultimately, great attention is given to large multiple death fires in public places such as hotels, nightclubs, and office buildings. But the major attention-getting fires that killed 10 or more people are a few in the number and have constituted only a small proportion of the overall fire deaths. Firefighters statistically are doing a good job in protecting public properties in this country. Furthermore, these properties are generally required by local fire prevention codes to have built-in fire suppression systems. The area with the largest problem is where it is least suspected, in people's homes. More fire prevention efforts must be focused on this part of the overall fire problem in order to prevent the majority of fire deaths. Only 3 percent of the 2001 fire deaths occurred in commercial or public properties. Outside and other miscellaneous fires, including wildfires, were also a small factor, 4 percent combined, in fire deaths. With frightful similarity fire injuries match fire deaths as far as place of occurrence, with 73 percent of all injuries occurring in residences. Fire injuries are distributed to other property types as nonresidential structures, 10 percent; vehicles, 8 percent; and outside and other fires, 8 percent. The picture changes somewhat for dollar loss. While residential structures are the leading property for dollar loss, nonresidential structures play a considerable role. These two property types account for 82 percent of all financial losses. The proportion of dollar losses from outside fires may be understated because the destruction of trees, grants, etc., is often given zero value in fire reports if it is not commercial cropland or timber. As a final note on property types, structures being residential and nonresidential account for only one third of fires, but they account for more than 80% of fire deaths, injuries, and property losses. Fire prevention efforts must continue to focus on home safety.[8]

The fire problem by property type remains a steady constant. In terms of numbers of fires, the proportion of the problem due to non-residential structures and vehicles has declined slightly over a 10 year period. The proportion of miscellaneous fires, although small has more than doubled over 10 years, likely the result of a small data set. Residential structures and outside fires have been relatively constant with a trend increase. Over a 10-year period, residential property fires caused an average of 73 percent of total fire deaths with sharp increases in the year 2000 and 2001. The trend in the proportion of vehicle fire deaths has risen slightly by 4 percent. Non-residential structural fires represent a small, but decreasing, proportion of deaths. The proportion of outside and other fires has declined substantially as well. Moreover, the proportion of injuries in residential properties and other fires has increased six and 9 percent, respectively; injury trends have decreased for the other three property types. The 10-year trends in the proportion of residential structure fire at dollar losses has increased 12 percent, reversing the previous downward trend, while the proportion of non-residential dollar losses decreased 19 percent. Vehicle losses have continued to trend upward, currently at 20 percent. Dollar losses in the other property type categories increased by 21 percent over the 10-year period. This jump is due to a 300 million loss in 1992 and a 390 million loss in 1998, both the result of large forestry fires. Residential fires have the highest number of deaths and injuries per fire and are another important reason for prevention programs to focus on fire safety. Non-residential structural fires have by far the highest dollar loss per fire.

Through statistical research we can profile the major causes of fire. In the year 2001 the profile of major causes of fires, fire deaths and injuries, and direct dollar losses for all property types grouped together is as follows. At 25 percent, incendiary and suspicious fires are the leading cause of fires. Cooking fires cause another 15 percent. The two leading causes of civilian deaths are smoking and arson at 23 percent each. These percentages are adjusted, which proportionately spreads the unknowns over the other 12 causes. The leading cause of injuries is cooking with 23 percent, followed by open flame 14 percent and arson at 13 percent. The causes of fire deaths and injuries are similar for both male and females. Regarding deaths, the most notable differences are in open flame and other heated fires, where male deaths are one third greater than female deaths; and secondly in electrical distribution fires, where female deaths are 40 percent higher than male deaths. A higher proportion of men are injured in arson fires and a higher proportion of women or injured in cooking fires. Similarly, for men and women, the two leading causes, arson and smoking, account for 45 percent of fire deaths. The two leading causes of injuries which are cooking and open flame, account for 43 percent of female injuries and only 35 percent of male injuries.[9]

Progressive Movement of the Study of "Fire Related Human Behavior" in all Its Applicable Aspects

The study of "Fire Related Human Behavior" is an essential part of the total, systematic assault on the fire problem in the United States. Negatively, there are human factors in conditions of unwanted fire, either through intent or carelessness. Of significance, the prevalence of arson and the improper use of cigarettes and alcohol. Inappropriate conduct through actions which reduce the effectiveness of the installed fire safety systems also plays a role. On the positive side human capabilities can be utilized, like those of a fire protection engineer or fire protection specialist, to a degree to enhance installed fire safety systems or compensate for the shortcomings in such systems.

The study of human behavior as it relates to fires is still in its early beginnings. It remains that this area of scientific study is beset by mythological problems. Of significance, experimental subjects obviously cannot be placed in real fire situations and after-the-fact testimony from participants in fires often contains errors and were samples are limited or not representative, conclusions must be drawn cautiously. Still, there is considerable consensus among scientists concerning certain recurrent themes which have emerged from these studies. Ultimately, the central perspective on human behavior in fires can be stated in the following. Despite the highly stressful environment, people generally respond to emergencies in a rational, often altruistic manner, in so far as is possible within the constraints imposed on their knowledge, perception, and actions by the effects of the fire. Simply put, instinctive, panic type reactions are not the normal occurrence at fires. The historic emphasis in the news media describing panic reactions is inaccurate and quite often counterproductive. The professional "Fire Related Human Behavior" researcher will cringe at such dramatic headlines in the popular press as panic kills fire victims. Realistically, when confronted by fire people will react through training instilled by routine. Today's researchers turn their attention to the whole range of behaviors exhibited by persons in fires, from the time they are first made aware of a possible problem through their completion of the evacuation process. The majority of data is often obtained through studies of actual fire incidents. In-depth case studies of individual incidents are conducted as well as statistical summery studies of large numbers of similar incidents. Today's researcher has progressed from largely descriptive studies like what was done, how often, by whom, in what order, to a more complex analytical study which attempts to extract typical behavior patterns or relate both behavior and fire development in a time sequence. Professor Bryan, the founding father of "Fire Related Human Behavior" studies in America, used this technique in an extensive series of case studies of healthcare facility fires. Relating behavioral actions to a timeframe is particularly important, since the appropriateness of an act such as fighting a structural fire is critically related to the stage of fire development in which it occurs. The entire behavioral process is taking place as the fire itself may be rapidly developing and what is considered an inappropriate action at one stage may be in fact appropriate a minute later. A further direction which many researchers are now taking, which was suggested by Bryan, is the exploration of the reasoning and motivation behind a potential fire victim's choice from among the alternative actions available to him. Why did he or she choose to perform a given act first and then did the outcome of this action match his or her expectations? Moreover, much attention is now being directed towards identifying the sequence of actions followed by those involved in fires. Unfortunately, it is not as simple to prescribe the appropriate sequence of actions to follow in a fire incident, as it is to specify a step-by-step standard operating procedure as, for example, in the assembly of a bookcase. Unfortunately, no two fires are alike and there are no official strategies for life safety that can stipulate a correct sequence of actions to follow. Moreover, the physical environment cannot be designed to literally reflect a specific sequence of actions.

Nonetheless, there are critical important directors which are generally applicable to fire incidents. For example, response to alarms should be immediate and appropriate, without time being wasted seeking verification of the existence of an actual fire, the occupant should not engage in fire suppression where there exists a moderate fire condition. Instead, the door to the room of the fire origin should be closed immediately after all persons have left; and rooms or buildings should not be reentered during the course of a fire to retrieve possessions. In healthcare facilities, frequently more than one staff member is close by, so that while one staff member is evacuating the immediately endangered patient another staff member can be pulling the fire alarm, and closing the door. Ideally, automatic detection devices should be in place to automatically sound the alarm at the earliest signs of smoke and heat.

The Study of "Fire Related Human Behavior" is responsible for current descriptions as well as the many detailed analysis of recommended evacuation processes. The initial physical science "carrying capacity" approach to egress assumes that occupants respond immediately to an emergency for example smelling smoke and are affected only by spatial configuration and density during an actual evacuation. The "human response" school of research considers the influence on evacuation time of such human instincts as design making in an ambiguous situation or a workplace factors such as the presence of trained supervisors or a public address system. Furthermore, the field of "Fire Related Human Behavior" is responsible for computer simulations of egress behavior in building fires as they incorporate human interaction and decision making variables. Behavioral research of fire incidents has shown that it is erroneous to assume that behavior in a fire situation is a simple process largely controlled by exits and alarm systems.

Notable researchers have observed that large-scale evacuations of high-rise office buildings that an exit that is not normally used will carry significantly fewer people in an actual fire evacuation. Fire prevention codes currently credit exit capacity to stairs regardless of their normal use. Researchers in the field assert that even in simple, total evacuation drills, evacuation times have been observed to be as much as twice as long as had previously been predicted in giving improved awareness of the complexity of behavior in fires, we should think of evacuation time predictions, even those based on realistic conservative flow assumptions as minimums and not maximum, as is sometimes disputed. A study revealed in one nursing home fire, perimeter stares not normally used by residents were not used by staff to evacuate residents in a fire incident, even though their use in the emergency would have been most appropriate. Subsequently, a center stairway was used, although it meant that most residents were moving towards the fire. A security alarm had been set off often in the past by residents who had attempted to leave through the perimeter exits and some residents who were caught using these exits had been reprimanded. One possible explanation for the impulse to use habitual routes during emergency evacuations may be found in the behavioral research regarding fear and stress. It has been proven that in heightened anxiety and fear people's attention becomes narrowly focused; they are aware of only the most obvious aspects of their environment. Peripheral information, which is usually easily processed, remains unseen.

The question of the effectiveness of alarms is dealt with frequently in the study of "Fire Related Human Behavior;" either directly or exclusively as an accounts of experimental studies or as part of fire incident case studies. Whether or not a fire alarm will wake a person is dependent upon more than just one noise level or its noise level in relation to ambient noise; sounds that have meaning to a sleeping person are more likely to wake him or her. Not only do individuals differ from one another as to the noise level necessary for wakening, but a given person will require different levels at different times depending upon such conditions as their sleep stage, time of night, or if they're on medication. Of significance, the frequency of false alarms, the possible ambiguity of the meaning of the alarm, and the tendency for people to look for confirming evidence of an actual file rather than to immediately evacuate, or factors which tend to determine the effectiveness of a fire alarm in producing prompt evacuation.

Those in the field of "Fire Related Human Behavior" have come to a general conclusion with regard to alarm systems of high-rise occupancies. Subsequently, the general consensus among researchers in the field is that a simple alarm system in a high-rise occupancy is inadequate. Moreover, total evacuation within a reasonable amount of time is not a feasible option. The question is therefore raised, how willing will people be to wait their turn to be evacuated or remain in a safe area of refuge once they are aware of the existence of a fire incident? The fire in the World Trade Center on April 17, 1975 demonstrated that even though reassurances are provided to a public address system, it may still be difficult to convince people they are not in any danger when they see an obvious problem such as smoke. In this incident, a small trash fire in the fifth floor resulted in the evacuation of the 9th through the 22nd floor because of occupants concern over smoke. Of significance, the fire safety director had initially urged people to return to their offices; when it became obvious they were not going to do so, he basically was bullied into ordering the evacuation. One aspect of partial, selective evacuations which could prove difficult is the condition where people working on the floor above a fire might be asked to go up a flight to prevent clogging narrow stairwells leading from the fire floor. Expectantly, most people would want to go down the stairs and leave the building. However, Professor of Fire Science for John Jay College Norman Groner found in several buildings that there were no instances during test drills when people fail to follow the somewhat counter instructive direction to go up stairways, although several occupants were to question the sense of this directive ultimately, the occupants could not be certain that this was only a test. In large buildings without voice communications systems, occupants may be forced themselves to choose between using their apartments or office units as areas of refuge and attempting to evacuate. A recent trend has been to provide increased access to public buildings to the physically handicapped. The elevators used to provide access for these individuals under normal conditions commonly are not designated to be used in the event of a fire incident. Subsequently, there are increasing recommendations to establish adequate areas of refuge in these buildings. It will be necessary to inspire confidence in the adequacy of such areas of refuge.[10]

The film and screen industry can be cited as the reason why panic is such a popular notion as to the course of fire deaths. Additional theories suggest that when the failures of others threaten the stability and predictability of our own world, we try to distance those failures from ourselves. Subsequently, we tend to dismiss accidental fire deaths as the victims fault; they panicked, but we would not, we often want to believe that the dead or injured were victims of their own maladaptive or panic stricken behavior. Of significance, no word has been used more commonly than panic to refer to "Fire Related Human Behavior." Additionally, no other word is more tantalizing or frustrating to a "Fire Related Human Behavior" researcher. The problem lies in the discussion of the meaning of the word Panic. Panic is often used at various times to refer to emotional action; to flight behavior alone, or to jumping are immobility reactions; to the initial reaction of an individual to a fire situation, or to the stampeding of a crowd as smoke and flames rush into a room with inadequate or blocked exits. Panic can refer to actions which might be labeled maladaptive or adaptive, rational or irrational, depending upon the outcome of the actions or whose survival is being considered. Can we see the concept of panic as a useful one for scientific purposes? Most scientists feel that the concept of panic is troublesome and irrelevant but its use in escapable due to the long tradition of use and its frequent appearance in the film and screen as well as news accounts. Most recently behavioral scientists have taken the nonscientific word like panic and have attempted to attach an operational definition to it for study purposes. Subsequently, even here there are intricacies as, when one attempts to attach modifying words, like rational or manipulative to actions like flight. What may seem like a rational behavior from the point of view of an outside observer may in fact be rational behavior from the point of view of the participant. What is maladaptive for the welfare of a group may be adaptive for a given individual. Movement within a burning building against the rush of the exiting crowd by a father or mother attempting to find a child is adapted from the parents viewpoint, but not for the group of people trying to get out.

Should the behavior of four occupants who recently jumped from a third floor window of a burning Brooklyn apartment building, receiving serious injuries, be labeled as panic? Noteworthy, is that all of these individuals before jumping had experienced unsuccessful evacuation attempts, receiving burns in the process.

The Beverly Hills Supper Club fire in Southgate, Kentucky, on May 28, 1977, in which 165 people died, is a case where a superficial look might lead one to ascribe many fatalities to panic. A British newspaper reporter at the time reported this fire with the headline "panic kills 300." In direct contrast to this, behavioral researchers agreed that panic, in the sense of aggressive behavior which would add to the danger to him or her and others presented by the fire itself, did not occur. Of significance, it was concluded by the research that there was an abundance of altruistic behavior. Moreover, a central behavioral problem laid in the failure to appreciate the seriousness of the situation; there was a false sense of security because the cabaret room where most of the patrons were located was a long way from the room of fire origin. The lesson learned hence, was fire safety education should now consider people's erroneous conceptions about distance being related to safety and the time needed to escape a fire emergency. Additionally, some patrons initially regarded the busboys announcement of fire from the stage as part of a comedy routine. Research of this fire shows that people need sufficient information about fire before they can, or are prepared to, leave the building.[11]

Growing evidence suggests that the delay in warning people in a number of major fires has been the primary reason why people have been unable to escape in time. An emphasis on avoiding panic contributes to delays. Any hesitation to inform people of a potentially dangerous situation out of fear of causing panic most often leads to a situation in which a panic inevitably occurs. Fire Related Human Behavior researchers are now focusing on what they consider the reverse of panic, known as inaction. Today's focus of concern is on inaction, denial, or the fear of appearing foolish by overreacting, the need to investigate before leaving a burning building, reentry of it, or persistence in fighting a fire too large to control rather than promptly leaving. Many studies are now focusing around revealed reentry behavior. Many people reenter an apartment building to get possessions or pets after reaching safety even after having seen smoke and flames while outside in the street. Classic experimental studies indicate that the presence of others lessens the chance of an individual reacting promptly to a potential emergency. In one behavioral study male undergraduate subjects found themselves in a smoke-filled room. When alone, 75% of the subjects reported the smoke. In the presence of two non-reacting others only 10% of the subjects reported the smoke during the experimental.

One strong pattern identified by a research is the tendency of people in fires to do the familiar; they use familiar exits, and they assume familiar roles. Findings from the behavioral study of the "Beverly Hills Fire" found that people involved continue to fulfill the roles they had prior to the fire. In that fire, the staff consistently took actions to assist patrons, whereas patrons followed or took a more passive role. Staff members took care of the patrons they would normally serve. Thus the conclusion from the research was that fire safety plans for places of public assembly should examine the roles that people normally play and not seek to prescribe emergency actions that are contrary to these roles. Another extensive study of nursing home fire incidents showed that nursing staff performed in accordance with their roles of responsibility for patients even when at some risk to themselves. Behavior in accordance with traditional male and female roles has been identified by several researchers. For example, a University of Surrey study found that the ability of the actions which follow the encountering of smoke and fire its self is explained by males and females differently. Females are more likely to warn others and wait for further instruction. Alternatively, they will close the door to the room of fire origin and leave the house. In both cases, females are more likely to seek assistance from neighbors. Male occupants are most likely to attempt to fight the fire. Male neighbors are more likely to search for people in smoke and attempt a rescue. The Wood study revealed sex differences, women were again more likely to warn others and evacuate the family while men were more likely to attempt to fight the fire. Conventional sexual behavior was also found in the Bryan, Keating and Loftus, Kobayashi and Horiuchi studies. Another strong behavioral pattern identified is the tendency of people to seek verification of fire clues before evacuating. There is also the problem of non-response due to negative conditioning by false alarms.[12]

People involved in fires often face difficult decisions because of the very limited time available in which to decide on a course of action. Moreover, decisions may be intellectually different in the context of limited knowledge of the engineered safety or of the basic configuration of the occupied structure or limited knowledge of the developmental stage of the fire itself. Decisions may be difficult because of the sometimes counter instinctive nature of the correct response; because some additional risk to oneself is incurred by a decision to alert or assist others. Furthermore, matters are complicated by the possible negative psychological effects of toxic gases or oxygen deprivation even before these factors produce severe physical symptoms.

The famous Illinois nursing home behavioral study found staff was resistant to accept the concept that ambulatory patients should be evacuated first. A group which received training on the home's fire emergency plan scored significantly better than a control group on test items regarding simple factual information relevant to fire safety, but no significant difference was found between the training and non-training groups with regard to the question of whether to evacuate amatory or non-amatory patients first. A majority of both groups incorrectly indicated that ambulatory patients should be evacuated first. The researchers concluded that, in instances like this, where there are strong erroneous preconceptions regarding appropriate behavior in fire, more thorough training methods, including simple explanations of why such beliefs are in error, are needed.

A fire that struck at a nursing home in Ontario Canada resulted in 25 patient deaths and a nurse was therefore faced with the problem of trying to save a patient in the room of fire origin. Records indicate when the nurse arrived at the fire room several items including the bed and chair were on fire. She could not enter the room, but was aware of the patients still in there, lying on the floor moaning in pain. She closed the door and reportedly went to get wet sheets and blankets in an attempt to save the patient. A supervisor returned to the room, opened the door, but unfortunately was unable to close it again due to the great volume of fire and smoke.

It is clear that the degree of altruism and capability exhibited by people in some of the case studies, studied by Fire Related Human Behavior researchers is perhaps surprising. Some cases in point exhibit man's humanity toward man for example actuation assistance rendered by a resident of another nursing home to another blind resident and the instance of occupants assisting other less able occupants in a New York City high-rise residential building.

A New Change in Direction Driven By Research

The reactive impulse of "panic" has been singled out for special attention by behavioral researchers. Whether or not panic occurred in a given situation to some extent seems to depend upon the definition one assigns to panic. Of significance, the concept of panic is too prevalent and two important to be dismissed merely as a semantic matter. Panic is no doubt the behavioral most commonly associated with fire incidents by the public in general and historically by the film and screen industry as well as those with professional interests and fire safety. Institutional fire safety plans and general safety guidelines to the public routinely and repeatedly warned against reacting with panic.

Fire Related Human Behavior researchers now tend to discount the importance of panic as a factor which adds to the injuries or loss of life which would result from the fire itself. Research conducted by the author of this book indicates that panic behavior is an assumption, not a condition to be defined or verified. Moreover, it is important to distinguish between panic, as an emotion, and panic, as an action with aggressive, maladaptive, or irrational makeup. People naturally will feel strong anxiety in a fire and emergency but this strong notion does not necessarily preclude appropriate life-saving actions. The consensus among Fire Related Human Behavior researchers is that the popular belief in the widespread prevalence of panic in fire disasters is a myth. Of significance, some caution is perhaps in order, when generalizing this downgrading of the importance of panic reactions in disasters to fire incidents. Mainly because of the extremely limited time for reaction and the possibility of sensory deprivation, fires may present a special category of disaster. The authors believe is that evidence of panic reactions should not continue to be looked for by fire investigators and researchers, the author suggests that more attention to other potentially harmful behaviors such as failure to respond to alarms or reentry of a burning building will prove more useful.

Decision-making during the various stages of fired growth can present a severe challenge to the fire participant. Since every situation is somewhat different, successful coping with a fire incident can demand more of a participant than following a set of previously memorized correct actions. The author believes that analysis of the decision-making process, based on in-depth interviews with participants, is one direction which might be emphasized in future research.

An important goal of fire related human behavior research is to understand the capabilities of people in fires and to be able to predict their likely reactions in emergencies. The extent to which human behavior is predictable is still an open question. Since human response in a fire incident tends to be difficult to predict it can be argued that an ideal system of fire safety might be one in which there is no dependence on human response and the approach could be classified as idiot proof. Practically, the author of this book believes there is no safety system that cannot be enhanced by appropriate human reactions or defeated by inappropriate human action. To some degree, human response can be compensated for shortcomings in the physical safety of the environment and vice versa. Of significance, each has an area of participation in overall safety that is not interchangeable with the other. An appropriate balance of all fire safety, both physical installation and human response in nature, is needed. In this balanced confinement, are factors including potential safety, reliability, failure mode in affect, cost and other impacts need consideration. Ultimately, absolute safety against fire injury and death is unattainable. Fire departments must implicitly accept a certain reasonable level of risk when they make decisions in the area of resource allocation for fire prevention and suppression.

With regard to the populations that are vulnerable the aged, the physically handicapped, and the mentally impaired; the acceptance of a certain degree of risk is done with the intent to improve their quality of life. Such individuals have been moved from institutional settings, with great built-in fire protection, to small residences with more homelike atmospheres. Greater access to public buildings has been provided for the handicapped ramps and elevators, placing them at greater risk from fire, but improving their freedom of movement and their quality of life. It is the authors believe that the handicap should be allowed to take the kinds of risk we all accept as part of normal life, since if all elements of risk are denied handicapped persons, then essentially developmental opportunities are also denied. The author also contends that the handicap should not have to endure more increased risk than those who are not disabled, and that some special provisions will therefore have to be made for the handicapped. It remains that determination of the appropriate share of the limited funds available for social concerns be devote to life safety from fire in consideration of the proper balance between reasonable safety from fire and quality of life are to of the complex, underlining philosophical challenges which face today's Fire Protection Specialist.

Chapter One Summary

The National Fire Data Center has a wealth of publications available free of charge that address the fire problem in United States. In addition to ordering through the online catalog, publications may be ordered by calling the publications Center at area code 800-561-3356 between 7:30 a.m. and 5 p.m. Eastern standard Time. To order publications by mail write to the publications Center United States Fire Administration 16825 South Seaton Avenue Emmitsburg, MD 21727 and please include your name, mailing address, daytime phone number, date required, title of the publication, and the quantity you need when ordering by phone or mail.

Two United States Fire Administration reports that have attracted nationwide attention are "*America burning*" and "*Fire death rate trends: an international perspective.*" *America burning* is probably the most widely quoted fire protection publication. This report sets the stage for national consciousness rising about the need for as much focus on fire prevention as on fire suppression. *Fire death rate trends: an international perspective* explores the magnitude and the nature of the fire deaths problem in the United States. It provides a statistical portrait of fire death rates in 14 industrialized nations and presents observations about key institutional and attitudinal differences between the United States and industrialized countries with significantly lower fire death rates. The United States Fire Administration website http://www.usfa.fema.gov also offers a wide variety of information on fire related issues. This site also features sections on fire related press releases, the National Fire Academy, a data center, fire safety, a kid's page, wildfire information, arson prevention, facts on fire, and the National Fire Incident Reporting System known as NFIRS.

This first chapter has served to acquaint you the reader with a broad range of research done in the area of human behavior in fires, it gives an overview of human response patterns for various fire safety design features and whenever possible points out some areas where various researchers have arrived at similar findings regarding behavioral tendencies in response to fire emergencies.

Discussion and Review

1. Is the fire problem much larger than generally known?

2. Does the United States have a fire death rate equal, less than or greater to several European nations and what is America's overall percentage; when compared to most European Countries?

3. What is the estimated total cost of fire to society?

4. What is the cost percentage of fire in relationship to the gross domestic product?

5. What is the perception of fire in the United States?

6. Which states continue to have the lowest death rates among the high population states?

7. What part of the United States continues to have the highest fire death rate in the nation and one of the highest in the world?

8. Simply put who suffers from the bulk of fire deaths?

9. The focus for injury prevention from fire should be on adults of what age group and why?

10. Will fire safety programs aimed only at the highest risk groups reach the majority of potential victims?

11. Is there a higher fire problem for some ethnic groups and why is it helpful to identify these groups?

12. In what type dwelling does the largest percentage of fire deaths occurs?

13. What are the two leading causes of civilian deaths by fire?

14. What are the leading causes of fires?

15. What have notable researchers observed at large-scale evacuations of high-rise office buildings?

16. Why is relating behavioral actions to a timeframe particularly important?

17. What do classic experimental studies indicate when there is a presence of others?

18. Any hesitation to inform people of a potentially dangerous situation out of fear of causing panic most often leads to?

19. What is the author's believe with regard to the evidence of panic reactions?

20. What does the author believe is the one direction which might be emphasized in future research?

2

Trusting Research and Putting Faith in What It Tells Us

o o o o o o o o o o o o o o o o o o o o o o o o o o o o o o o o

"Hardware cannot rule out the human element from any emergency situation, and training cannot guarantee appropriate behavior. However, designing procedures and equipment around human behavior rather than trying to shape human behavior to fit procedures and equipment can compensate for those limitations."

—*Groner, Loftus, Keating in*
American Hospital Publishing

Introduction

In the not so distant past, medically practicing psychologists working at a Staten Island, New York psychiatric Hospital known as Willow brook engaged in "aggressive behavioral modification." Historical records indicate that doctors actually engaged in cruel and unusual physical abuse and may have actually caused the death of some of their patients. Subsequently, and only as a result of the discovered maltreatment of the mentally handicapped have behavioral scientists and doctors working in the field of psychology taking a more scientific approach to medical theories and techniques with regard to treating the mentally afflicted.

Presently the "Scientific Behavioral Method" has gained tremendous popularity and has been effective in improving our standards of living. Of significance, we have come to trust research, and we tend to place an incredible amount of faith and what research comes to tell us. What separates humans from other species is our profound awareness of our own mortality. Subsequently, our preoccupation with death leads many of us to try almost anything to add years to our lives. When we combine the enormous trust we place in science with the strong desire to live longer, the result is a society that often fails to adequately question theories, speculations, and claims that seem to provide us with explanations and answers.

Techniques That Appeal To Emotion Instead Of Reason

We trust science because we want life to be predictable. When we encounter a situation that is unpredictable, we seek to understand it, explain it, and ultimately control it and predict our outcome. Ultimately, these are the goals of the Human Behavioral Scientists. Subsequently, it should not seem unusual for people to have a willingness to accept as factual scientific theories those that attempt to explain that which is difficult to understand. Moreover, our legal system has also advanced the position that scientific evidence ultimately will lead to the truth. You can't scan through your television broadcasting without coming upon a police or courtroom drama that deals with subject matter of both forensic and witness account material. In the Criminal Justice System lawyers and others can use science, in the form of forensic evidence and expert witnesses, to either substantiate or rebuff almost any theory or position. This being the case, we often ask ourselves how I differentiate between valid and invalid scientific evidence. In the legal system, it is often the persuasiveness of the presenting attorney, not the validity of the evidence being presented, that often convinces a jury. However, from a scientific standpoint the rigor of the experimental design, the statistical significance of the results, and successful replication of the findings all affect persuasiveness. The author's goal for the reader is that by the end of this chapter he or she will be able to understand how scientists can evaluate each of these factors. But for now try to understand that you must use a scientific approach to evaluating research and not merely the look of the presentation.

How convincing is a chart or graph to you? The following picture shows a 25 percent increase in the number of cases of suspicious fires reported from the year 2002 through 2005. Since we are all sensitive to the word suspicious, the picture has even more dramatic affect.

Number of _"Suspicious Fires"_ in the Borough of Brooklyn, New York City

2002   =   720

2003   =   741

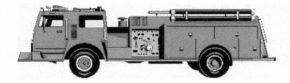

2004   =   759

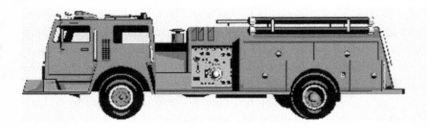

2005   =   802

You should have notice that while the numbers between 2002 and 2005 represent a 25 percent increase in the number of suspicious fires, the picture represents a 300 percent increase in the number of fires. For the picture to be an accurate representation, the largest piece of fire apparatus should be about one quarter larger than the smallest fire engine, not more than three times the size, as the pictured chart depicts. A pictured chart should help you understand the data. If it appears to be trying to convince you, then you should be leery of its intent. Additionally, the word suspicious is often confused with an official proclamation of arson. Any fire chief can consider a fire to be suspicious merely because its origin is not readily identifiable. Moreover, only after a complete and thorough investigation that entails a review of laboratory deciphered physical evidence as well as the completion of witness interviews can a fire be officially declared arson by a fire marshal. Of significance, most initially reported suspicious fires are declared not to be arson by a fire marshal and are the results of a non-malicious act.

With regard to scientific data, many factors contribute to misrepresentation and some are more important than others. Ultimately, you will want to keep these factors in mind when analyzing data. When you read a newspaper article, you might notice that the reporter seems to be trying to convince you of his or her position. Though research is supposed to be an objective analysis, reporters and their organizations often have strong feelings in their positions concerning the topic. Subsequently, in their eagerness to present a convincing argument, writers often, intentionally used a variety of persuasive techniques. Generally, these techniques appeal to emotions rather than to logic and reason. The author has composed a list of a few of the most widely used persuasive techniques. The "name-calling device," plays upon the hatred and fee is of the reader by labeling the source with bad names to draw attention away from evidence. The second technique known as "glittering generalities" appeals to emotions through using good words to draw attention away from bad ideas and evidence. The "testimonial technique" asks the reader to accept uncritically a viewpoint because someone important espouses the same position. The "card stacking" technique uses half truths, false evidence, selective citing of evidence, and distortion to confused the use and to make the reader accept the writer's viewpoint. Subsequently, "card stacking" presents only evidence that supports one view when opposing evidence exists. The last of the most widely used persuasive techniques is known as the "bandwagon." With regard to this persuasive technique the reader is asked to accept a viewpoint uncritically because everyone else has done so. You should be aware that writers as well as researchers who used these types of techniques seldom employee only one of them. You probably can recall reading statements similar to the following sentence. "Experts well-versed in the subject agreed that all conclusive evidence indicates that." The preceding statement includes glittering generalities (the words well-versed in conclusive), testimonial (the opinion of experts), card stacking (no opposing view is presented), and bandwagon (the statement strongly suggest that all reasonable people will accept this truth).

When an emotional story is presented as hard proof of some common casual relationship, we, as intelligent consumers of information, should regard the proof with skepticism. Although these stories may be moving, they lack one important element that we must demand before we can accept them as a universal truth. Of significance, to be recognized as a universal truth a finding must be true for a large segment of the population. For example, there are many anecdotal stories regarding teenagers who commit violent acts because of their participation in violent video games. While we cannot draw the conclusion that playing such video games causes people in general to commit violent acts, there are few cases in which playing certain video games has caused a few teenagers to imitate that behavior. Since we also know that imitating the behavior of others is a powerful factor in influencing behavior, there is a practical significance related to the anecdotal evidence described here. Even though we know there is not a significant casual relationship, we are duly apprehensive about allowing teenagers to play violent video games. Hence, video games are rated in order to give parents an idea of what type of material their children may be viewing.

From a fire prevention perspective, most people know that the use of candles is dangerous in a practical significance. Statistically, candles pose little threat to the general population, but most people are still careful when using candles. Thus, because people are aware of the practical significance of the danger of an open flame, they are cautious; therefore, injuries from candles are not statistically significant.

You may have described the idea of a high percentage. If your definition has something measurable about it, then you are thinking about statistical significance. On the other hand, if you said that significance had to do with something such as importance, then you are thinking about practical significance. While both types of significance are important and useful, most behavioral scientists rely on statistical significance. Quite often in behavioral studies you will see statements such as "95 percent of the people expressed having cleaner clothes." Although this statement may be true, we still need to answer some questions before accepting the statement as evidence of the products capabilities. You should be asking questions such as, "cleaner than what?" After all, almost anything would be an improvement to not washing your filthy clothes at all. You would probably want to know with what the clothes were soiled. You also might question the method for measuring cleanliness. For the remainder of this chapter, let's just consider your question about making comparisons.

The Various Techniques Used To Study Human Behavior

The six most common techniques currently used by researchers to study human behavior are case studies; surveys and questionnaires; experiments; correlation analysis; models as well as computer simulations and media analysis.

Case studies make up the oldest and some of the best known studies on human behavior. Researchers conducting a case study focus on the human behavioral aspect of a single individual or event. Add example of research in fire related human behavior using the case study method include the work of Best and Lawson, both researchers focused on behavioral aspects of the Beverly Hills Supper Club Fire, which occurred in Southgate, Kentucky, in 1977.

Typically, researchers to develop case studies use a range of available data, including interviews with victims and firefighters, witnesses, first-hand observation and official fire reports of the incident. The early case studies on fire related human behavior tended to be descriptive and journalistic. In contrast, recent case studies are more systematic and reflect increased attention to mythological rigor. Case studies can give rich detail regarding a fire incident. However, one to his advantage of most case studies is their focus on the unique aspects of each fire situation rather than the comparable elements across many fire incidents. Subsequently, it is difficult to draw conclusions about other fire incidents unless those emergencies are very similar to those in the case study.

Surveys often include either face-to-face interviews or self administered questionnaires. A survey attempts to measure something about a population by asking individuals a series of questions about the research area. Different aspects of fire related human behavior have been studied by means of surveys, including but not limited to attitudes and behavior with respect to fire hazards, the behavior of persons involved in fire emergencies, and the effectiveness of fire safety programs. Surveys are usually utilized in fire research because they make it possible to collect large amounts of uniform data across a single range of fire incidents. Subsequently, to be most useful surveys must be designed and administered properly. Of significance, those conducting surveys about victims who died in a fire situation should administer the surveys very soon after the event to minimize distortion and loss of information. They also must take care to administer the survey consistently to all of the participants. Since it is usually not particular to administer the survey to all members of the population being studied, it is important to ensure that the survey group is representative of all that population.[13]

Additionally, researchers used both laboratory and field experiments to understand human behavior. Moreover, the experimental method is a powerful research technique cool as it enables the researcher to have considerable control over the stimuli that subjects experience and to test specific hypotheses about human behavior. However, fire related human behavior experiments have difficulty realistically re-creating the fire situation and it is impossible to study some aspects of fire is experimentally. It remains unethical to expose subjects to actual smoke and flame to see how they would react. One experiment conducted by Latane and Darley, which led to generalizations about how the presence of others influence is an individual's response to smoke. The study indicated that the presence of other people inhibits some from recognizing ambiguous clues as indications of a real threat.

The Latane and Darley experiment was given to college students under the following circumstances. The students were asked to go into a room and fill out a questionnaire. Unbeknownst, to the students, it was not the questionnaire itself that was important to researchers but their response to the stimulus that was introduced into the room. Behavioral researchers injected smoke into the room through a small vent hole. For a dependent variable in the research experiment, that being the aspect of the subject's response that was being studied, was whether the students in the room recognized the danger and reported the smoke and if so, how long did it take them to report the smoke? The independent variable, being the factor that the experimenters varied in different experimental trials, was the size of the group in the room. Subsequently, what the experiment showed is that group size was related to willingness to act on the danger clue and report the smoke. In general, a larger group size placed in the room, resulted with less willing individuals to come forward and report a threat of smoke. Subjects alone in the room reported the smoke and 75 percent of the cases. When groups of three subjects were involved, the percentage of time subjects reported smoke dropped to 38 percent. Overall, 24 subjects took part in the experiment in eight person groups; of these 24, only one person reported smoke, representing a mere 4 percent of subjects in the study.[14]

Additionally, Latane and Darley injected what is known as Confederates into the research. The Confederates were instructed to remain passive and do nothing about the smoke, no matter how dense the smoke became. Subsequently, students who saw the Confederates doing nothing did not report the smoke, despite the fact that conditions in the room became quite irritable, persons from larger groups who did not report the smoke took longer to do so than those who were alone in the room. Even noticing the smoke seemed to take longer when other people will present. Those who did not act on the dangerous stimulus used a variety of reasons to justify their inaction they reported being unsure about what the smoke was, they reported rejecting the idea that it indicated a fire, choosing instead to identify it as steam, papers in the air conditioner, even smog.

The results of these findings suggest that when trying to determine whether ambiguous clues indicate a file or emergency, individuals often look to other people. If others interpret the situation as non-life-threatening, the tendency to respond with concern diminishes. The results of the experiment partially explain why people delay escaping, reporting fires, and engaging in other self rescue actions.

Simulations typically use data taken from empirical observations or the results of other research. Computers and mathematical models give the researcher the opportunity to take into account many variables that could contribute to a given outcome or process, for example fire spread and to manipulate large amounts of data rapidly. Model should be able to be applied in such a way as to incorporate human behaviors. Later in this textbook we will discuss computer models more comprehensively.

There are two other types of research techniques that you may encounter. The **meta-analysis** and the **literature review**. They are both similar in that both of them review the existing written material on a subject. In a literature review, the researcher attempts to find all of the relevant written material on the subject. The researcher then lists and describes these sources, often evaluating, summarizing, and sometimes drawing conclusions from the material. In a meta-analysis, the researcher conducts a more objective analysis of the existing research by attempting to quantify and compare the results of various studies. In the meta-analysis, for example, the researcher might compare the results of all of the research on socioeconomic factors and fire incidents by listing all of the factors from each study. The researchers then would conduct a factor analysis to identify which elements were significant in each of the studies, yielding a new set of socioeconomic factors that encompasses all of the other factors. These types of studies are useful when there are a number of marginally significant or even contradictory studies on a topic.

Human Behavior Research Design

When conducting Fire Related Behavioral studies, researchers employ the scientific method, testing propositions by systematic approach, which includes the development of a hypothesis. Simply stated, a hypothesis is a statement that tentatively expresses a relationship between variables. It is not much more than an educated guess or a hunch about something. However, to test a hypothesis experimentally you must state it clearly in terms that are observable and measurable. You might want to prove the hypothesis that frustration leads to carelessness, which causes accidents. To test this hypothesis experimentally, you must be able to define "frustration" and "accident" in observable and measurable terms. Additionally, you might have to differentiate a "careless" accident from other accidents.

For now, let's simply try to hypothesize that frustration causes accidents, how would you define "frustration" and "accidents?" You might create a situation in which you offer to pay a $200 prize to the first person to put together a puzzle then you keep interrupting those people who are trying to complete the puzzle. You can define an accident as every time a contestant drops a piece. You should be able to control the frustration, how many times to interrupt the subjects, and count the number of accidents. Moreover, you might substitute to term "unintentional" for the term "careless." This would not satisfy the requirement because it still does not allow you to measure and count the events. The descriptions that we have used for frustration and accident are called **operational definitions**. An operational definition may be far removed from the actual meaning of the term. It is used to represent abstract terms, such as frustration, in real-world terms by stating the exact procedures that will represent abstract concept. Abstract concepts, such as frustration, are called **hypothetical constructs**. They are those things, invisible in themselves, that we cannot see but whose existence we know about because of their consequences. **Of significance the term "motivation" is one of the most widely used hypothetical constructs.** Can you think of examples as abstract as the concept of motivation?

The first step in designing an experiment is to develop and hypothesis. The second step is to set up the experimental situation to control the input and measure the output, respectively called independent and dependent variables. A common concern for researchers is experimenter effect. Most people would expect that researchers, when conducting research on human behavior, would not tell subjects in an experiment about the experiments expected outcome or even what the researchers have designated the experiment to measure. Most researchers even keep the subjects unaware of the independent variables, as in the case of sugar pills given in a place of an actual medicine. When the subjects do not know whether they are in the test group or the control group, but the experimenter does, the experiment is referred to as a **single-blind experiment**. While most research employs single-blind experiments, researchers often forget about the affects that the experimenter may have on the outcome of the experiment.

There are many cases in which teachers unknowingly affected the outcome of research conducted in their classrooms. When instructors received information related to their student's ability before an experiment, they gave higher grades to students who, according to that information, had the higher abilities. Subsequently, they gave lowest scores to students they believe had lower abilities. While these findings have tremendous implications for educators, the point here is that people other than the subjects of the experiments can sway or bias their results. It remains, however, that someone must know the exact experimental conditions. Of significance, they should be the principal investigator, who should not attempt to ensure that experimenter bias does not affect the outcome and should not conduct the experimental procedures directly. Experiments conducted in such a manner that neither the subjects nor the experimenter's are aware of the actual conditions are called **double-blinded experiments**. In the case of the sugar pills or placebos, the researchers could make the experiment double blinded by having a computer randomly assigned subjects to either the control group or the experimental group. Assistance could distribute pills accordingly; only they, and not the experimenter's or the subjects, would know which group was given which pill. Subsequently, pre-existing beliefs and biases would influence neither the researchers nor the subjects' actions and observations.

As demonstrated in the "Hawthorn effect" often, the experiment itself causes unexpected results. While conducting research on lighting conditions at a General Electric plant in Hawthorn, Connecticut, the experimenter's encountered some unexpected results. The experiment is married to lighting, trying to identify the optimal lighting conditions for plant employees. They found that, during the experiment, the worker's production was better even in the very poor lighting conditions than it ever was before the experiment. Production apparently improved because the workers interpreted the presence of the experimenters and the experiment as an improvement. Consequently, they worked harder on the old the experimental conditions. Naturally, these results would have been different if an experimental control group had been used.

As a result of the "Hawthorn effect" a proper research design requires the inclusion of **control methods**. When conducting experiments, scientists usually assigned subjects randomly to at least two groups. One group is called the **experimental group** in the other group is called the **control group**. In the earlier example involving the statement that 95 percent of the people expressed having cleaner cloths, the experimental group would be the one that used the new detergent while the control group would be the group that received the old detergent. Of course, the experiment may have had only one group. In that case, you could not place any real confidence in the results. Let's look more closely at how the researchers in that study of cleaner cloths measure cleanliness. There are some very legitimate questions you should ask about the manner in which the researchers conducted the studies and take the measurements. For instance, you might find that the measure of cleanliness was based solely on having an independent evaluator look at before and after photographs of the clothes and judging which appeared cleaner. Although you might feel this **method of measurement** is somewhat subjective, this type of measurement is often the only choice available to behavioral researchers. When encountering a subjective measurement method, you should ask, "Could the researchers have used a more precise, less subjective system of measurement in their study?" After all, in considering the subjective evaluations of before and after photographs, you must not forget that the purpose of buying a clothing detergent with a better cleaner would be to improve your appearance to others.

Could the results of an experiment be purely coincidence? If you flip a quarter 10 times, you would expect it to come up heads about 50 percent of the time. However, if you flip that same quarter 10 times every day for a year, do you think that it would come up heads eight times on one of the days? If your answer is no, then you must not believe in playing the lottery. After all, is in playing the lottery hoping to be a million to one odd? If your answer is yes, then you understand that there is always a chance that result of an experiment may simply be the result of chance, and believing in chance, you might as well be a lottery player. This is why most behavioral researchers indicate whether their findings are **statistically significant** or not. The statement actually would be that the results are statistically significant at, for example at .05. This means that the chances are only five out of a hundred that the results were simply a coincidence, and therefore 95 out of a hundred that they actually prove the hypothesis the researcher was studying.

In the following chart, more the research whose results you believe has the highest statistical significance or less chance that they are coincidence.

Fire Behavior in a High-rise Building	Level of Significance
Project # 123	.01
Project # 111	.025
Project # 234	.05

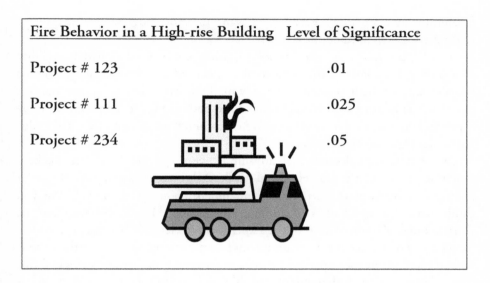

Since the chances are only one out of a hundred that the results and project number 123 were a coincidence, it would be the most conclusive. The rule is simple. The low of the number, the more coincidence you would have in the results. Behavioral researches with results that are statistically significant at the .001 level have one chance in a thousand that the results are coincidence. Of significance, most research is reported to have results between the .01 and .05 levels of significance.

Can you name the method that scientists and behavioral researcher's alike use to improve confidence in their research? The most common method used by scientists and behavioral researchers alike to improve confidence in their research is done by replicating the results. To replicate an experiment means simply to repeat the experiment; to replicate the results means to repeat the experiment and obtain the same results. For example, researchers trying to evaluate the ability of subjects to wake up within a given amount of time when exposed to the smell of smoke would conduct an experiment numerous times under similar conditions. The idea here is that if the research is valid, different researchers who replicate the experiment will obtain similar results. The word replicate is used instead of duplicate because the term duplicate implies obtaining the exact results, which would be highly unlikely. Often, when researchers replicate an experiment, their results differ greatly from those of the original experiment. While the original researchers may dispute that the experimental conditions were different in the replication, it still remains that the original findings are less valid in light of the new findings.

Therefore, in evaluating research you want to know how many times other researchers have replicated the results.

In collegiate terms the word validity signifies correctness! Generally, validity means that the results of the behavioral research all correct; that is, the results actually support the hypothesis. The only method that can even suggest that one thing was as another is the experimental method. However, although only the experimental method can actually prove causation, poor design often will evaluate the results of an experiment.

Correlation is the degree of relationship between two variables. When behavioral researchers use the **Correlational Method** they are measuring the degree of relationship between the dependent and independent variables. Why do you think that most case studies concerning Fire Related Human Behavior use the Correlational Method instead of the Experimental Method? Because, the underlining concern is that the areas under study are not appropriate for the experimental method. In the experimental method, researchers identify a single factor or variable, called the **independent variable**, and manipulate it, by adding, deleting, or varying it, in one group called the **experimental group**; but hold its constant in another group called the **control group**. Researchers then measure and statistically analyze the results called the **dependent variable**.

The results measured in Chapter one of this textbook, that is, the dependent variables, were fired deaths and fire losses. Naturally, no one would conduct experiments that would cause people to lose their lives or property in a fire, although we could have an experiment that would save lives and property that fires otherwise would have claimed. Of significance, a similar ethical issue arises here, as in medical research. Who would be subject to greater fire risk by placing them in the control group rather than the experimental group? In medical research, the question is often, "Who will remain untreated by being in the control group as opposed to the experimental group?" We said that fired deaths and losses are the measured results in the study, but now we are considering changing something to reduce these fired deaths and losses. What interventions can you think of that might reduce fired deaths and losses? Being fire science students you probably have a long list of answers for the previous question. For now, we will deal with the simple answer or solution of installing residential sprinkles. This idea should work well with the experimental method. We could have volunteers install sprinkler systems in one group of private dwellings and not in another group. In the experimental group, assistances conduct maintenance semiannually. After a designated period of time, we would compare fire losses between the two groups. If the results were statistically significant, in that the test group had fewer fires than our control group, we would draw a scientific conclusion that installing residential sprinklers systems reduces fire losses.

This case study is actually more of a quad side experiment, but the results are still valuable. If we look back at ideas that we listed for lowering fire losses the question is posed; can we manipulate the variables in the same manner as we manipulate the presence of a sprinkler system? It should be obvious that we are not going to be able to change the variables. Since we are not able to change these variables, we cannot use the experimental method for this case study. We must use a Correlational Method when we aren't able to change the independent variable or variables.

Now we can answer our question, "why do you think behavioral researchers use the Correlational Method instead of the Experimental Method?" Of significance, the primary reason is that behavioral researchers are usually not able to manipulate the variables. Instead, they isolate the variables and then compared each one with fire losses. Upon finding a high correlation between a particular variable and fire losses, behavioral lists conclude that the variable was somehow related to fire losses. Chapter 1 of the textbook makes correlational comparisons of many social economic groups. Although the text offers many explanations and suggestions for future of events, it does not suggest causation. That is because the correlational method does not prove causation. If it did prove causation, city government might correlate the number of firefighters at a fire event with the number of fired deaths and conclude that the more firefighters called the more fire deaths. Obviously, this is not true, and it shows that correlation analysis may be correct with out being valid. Not enough can be said for common sense the value of the correlational method is that it helps to suggest future of events from past occurrences. When compared with other analysis, a correlational method is very useful for gaining an understanding of "Fire Related Human Behavior." We can reduce the number of incorrect hypotheses by viewing the various associations among fire incidents.

A recent study conducted by the New York City Fire Department Bureau of Fire Prevention found that there were 20 household fires with casualties in low income areas, and that 15 of the houses had in operable or no smoke detectors. Intuitively, it appears reasonable to conclude that there is a relationship between income level and having a working smoke detector. However, this is not a valid conclusion because the survey considered only the co-occurrence of events being low income plus no smoke detector. Incidents, in which events are not a co-occurrence in low income and smoked detector, and high income and no smoke detector, are also factors. Subsequently, before inferring a valid relationship it is necessary to take into consideration the number of times operable smoke detectors are found after casualty causing fires in higher income residents. I illustrate this point hypothetically with the following data matrix. Of significance, people tend to pay attention only to the lower right quadrant, which gives the number of low income households involved in injury causing fires that had no or any inoperable smoke detector. Moreover, the data shows only a very small association between income level and having in operable smoke protector. Furthermore, given the small sample size, this relationship is not even in the realm of statistical significance.

Hypothetical Data Matrix

	Above Medial Class	Below Medial Class
Working Smoke Detector	4	5
Inoperable or No Smoke Detector	10	15

Subsequently, it is also attempting to include that this data shows that not having in operable smoke detector causes injuries, since about 73 percent of the injury causing fires occurred in houses without operable smoke detectors. Although this makes intuitive sense and may well be true, the data alone does not support the claim. The only way to know whether such an association exists is to collect and analyze data that shows how many operable smoke detectors were found in similar fires that did not result in injuries.

Create your own smoke detector matrix by filling in the headings that describe the data you want to collect test this hypothesis.

7	?
23	?

As you can see, data from the previous example fill in only the first column of the table. Before you can draw any conclusions, it is also necessary to know the number of operable smoke detectors in fires that did not involve injuries. Your matrix should look like the following one.

	Structure Fire W/Injuries	Structure Fire W/O Injuries
Working Smoke Detector	7	?
Inoperable or No Smoke Detector	23	?

Most importantly, you should have noted that we based the above examples on smoke detectors with categorical data. This type of data measures something that either is or is not. It does not provide a range of measurements. In the above examples, we were simply looking at whether a smoke detector is operable or in operable or not present, which amounts to the same thing in the event of a structural fire. For this kind of data, we would not use a correlational analysis. Subsequently, in this case we probably would use a different method called a chi-square test instead. Since most of the research on "Fire Related Human Behavior" will use the **correlational method**, you need to understand this method to evaluate effectively the research you may encounter in your collegiate studies as well as your career advancement. The correlational method compares traits, behaviors, and events, seeking the degree of relationship among variables. Remember from chapter 1 that there are both positive and negative correlations.

As a review let's check to see how much you remember? Is the relationship between smoke detectors in the home and fire death rates a positive correlation or a negative correlation? As more households in the community use fire detectors, fire death rates decrease. Therefore, there is a negative correlation between the presence of smoke detectors in the home and fire death rates, because as the presence of smoke detectors increases, the rates of fired deaths decreases. If you are still on shore about this theory, you should review Chapter 1 of this text before continuing.

Behavioral researchers conduct correlational analysis by using one of a number of statistical procedures and formulas. Some of the most common methods are the Pearson Product Moment Correlation, the Spearman Rank Correlation Coefficient, and the Biserial Correlation Coefficient. In this textbook I will go into the statistical procedures and formulas involved in computing these correlations. Of significance, you will need to understand the results of these computations. Fortunately, the correlational procedures or yield the same range of results. The result of a correlational analysis is a value between minus one (-1) and plus one (+1). Simply, now we can see where the concept of either a positive or a negative correlation originates. A perfect negative correlation is a minus one while a perfect positive correlation is a plus one. You might ask yourself, what does a result of zero (0) indicate? As you might expect, it means that there is absolutely no relationship between the two variables. Subsequently, as the result approach zero, the relationship between the two variables decreases. Invariably, the results of a correlational analysis are presented as decimals since the occurrence of any other result is rather rare. Some statistical experts suggest that correlational coefficients between .30 and .70 either plus or minus suggest a moderate relationship. Simply put this means that coefficients above .70 suggest a strong relationship and those below .30 suggest a much weaker relationship. Although one may find this to be a useful statistical rule I recommend you consider size of the group, nature of the variables being studied, reliability of the data; and sampling methods used as additional factors.

With regard to group size, a high coefficient may appear in data analysis involving a small group but the results may be meaningless because of the possibility that they are anomalies associated with small samples. Suppose you want to sample the average sick time for each firefighter in a particular group assignment in a particular firehouse. If a firefighter in one of the groups had a serious medical problem and her sick time was two or three times higher than other firefighters, the group average would be unusually high. As you would expect, the average, likely would decrease if the size of the sample was increased. A .70 coefficient might not be statistically significant with a very small sample size, while a .30 coefficient might be statistically significant with a large sample size. Although not always possible, an absolute minimum group size of 30 is preferred in gathering statistics for correlational analysis.[15]

When you consider the nature of variables being studied you have to make more subjective decisions. An example of this would be, what if you found a correlation of .75 he doing the presence of smoke detectors in the home and the presence of a firefighter living in the home? Would you consider this a high correlation? In contrast, if you found a correlation of .40 between the presence of multiple smoke detectors in the home and a below middle-class level income for the household, would you deem this a low correlation?

Take some time from your studying now and write down variables that you feel are present in your community with a high correlation. The variables should be fire related.

_____

_____

_____

Look over the list you just made. For each item, think about the population for which you would gather data and the method you would use to gather them. Also, consider how accurate and reliable this data would be. Are you having any trouble making your decisions? You should be because these are the same problems that all research's face. So what are you saying about the data? Write down your impressions on collecting data.

_____

_____

_____

_____

Ultimately, you should have reached the conclusion that data is often difficult to obtain and is not always reliable! Fortunately since the research that we will be evaluating will be fire related. We are in a good position to determine the accuracy of the data used in the research. You should record your concerns about the data, just as you did in the previous exercise. Then, as you analyze the research, see if the researcher's procedures satisfy your concerns about the data. We use the term reliability loosely to describe the accuracy of data. However, in research the term reliability has a different meaning, while similar in research **reliability** emphasizes the importance of consistency. In this sense, both statistical significance and replication are associated with reliability and can be explained as the likelihood that, if the same study were repeated, the same conclusions but not the exact results could be expected. Another term used by behavioral researchers is the term **internal reliability** which is used to refer to the fact that different raters will agree when they collect the data. A good example of an internal reliability problem involves the National Fire Incident Reporting System or (NFIRS) data system, which collects data from fire departments to obtain regional and national statistics about fire origins and losses. To the extent that different firefighters investigating the same fire would respond the same way to questions, the data are reliable. To the extent that they might respond differently to the same questions about the same fire, the data are unreliable.

The contrast of reliability is validity. Where reliability refers to the consistency of the data in the research, **validity** refers to whether the research really measures what it is intended to measure. The validity is questionable to the extent that it is possible to explain the results in ways other than the ones argued by the behavioral researcher. Of significance, not observing the protocol of the double blind experiment is a threat to validity, not reliability because other explanations are the results is possible for example placebo or experimenter expectation effects. Subsequently, another important validity issue is whether research generalizes to different settings. An example might be the results of a study of juvenile arson in the United States may not be valid when applied to juveniles in other countries, or to different age groups within this country. Of significance, this is referred to as **external validity**.

Why is it necessary for researchers to obtain a sample that is representative of the entire population being studied? Because, although you may see some research that reports on the entire population of the United States most research is done on a much smaller scale. The population may be very large for example all humans, or very small three to four people. If the population is small enough, the researchers used the entire population for the research and there is no need for sampling. When the population is very large, however, it becomes necessary to use **sampling techniques** to achieve a representative sample of the entire population. The purpose of sampling is to ensure representative ness as well as to reduce the possibility of statistical anomalies associated with a small population. Of significance remember it is preferable to use a sample size of at least thirty. The most common method of sampling is known as **random sampling**. In random sampling, everyone in the population is afforded the same chance of being selected for the experiment. Odds are that such a sample will be representative, meaning it is large enough. A common method is to use the last digit of a universal number to select the population. For example, the last four digits of a Social Security number or the phone number have been used by many researchers to select random samples. Of significance, you cannot use the first digits of phone numbers and Social Security numbers because they relate to certain geographical areas. The problem is that if they are from the same geographical area, say, the northeastern United States, we have to modify our conclusion to say, "Research shows that people from the northeastern United States often are." In order to test the validity of conclusions, ask yourself these questions. For example are these cities, either individually or when combined, representative of the nation? Are these cities, either individually or when combined, representative of your community? The second question may be easier for you to answer and more relevant to your needs. Though you may be unable to make conclusions about national trends, applicable to your local area may be of considerable use to you. Since most of the research relating to the demographics of fire losses is, out of necessity, conducted in specific urban locations, you must learn to make judgments about the relevance of the data to your own locality. I recommend a note of caution! Although it would be a mistake to except research that was not relevant to your locality, it would be just as large a mistake to reject research that may be relevant. By mistakenly rejecting research findings, you could be overlooking valuable information that might help save lives and property from fire losses in your community.

How do behavioral researchers explain **causation?** After reviewing research and finding that there does appear to be a high correlation between two variables, you may feel compelled to say that variable "A" cause's variable "B" or that variable "B" causes variable "A." for example, much of research has been written that indicates that there is a high correlation between overcrowding and fire deaths. Would you say that overcrowding causes fires? The answer is no. Based on this research, you cannot say that overcrowding causes fires. What you can say is that there appears to be a significant relationship between overcrowding and increased incidence of fire deaths. On the other hand, you cannot say that overcrowding does not cause fires. In fact, much of this research that was written give strong evidence that the condition of overcrowding leads to increased incidences that result in fires. Although many people might agree with you that overcrowding does result in more fires and subsequently ask you what they can do about the fire problem, other people are going to ask you what proof due you have. At that point you would show them the Federal Emergency Management Agency Report number 170, titled *Social Economic Factors and the Incidence of Fire*, pointing to the table that clearly shows the correlation between overcrowding and fire. The doubters will respond that correlation does not prove causation, which is the point of this discussion.[16]

Of significance, it is important to remember that the value of the **correlational method** is that it helps to suggest future events from past occurrences! This is why correlations are important to the fire service. If you can predict something, it is then possible to try to prevent or alert others before it happens as well as plan a response. Above all, this is what firefighting is all about and must remain the **core competency** or functional area of expertise for the Nation's Fire Service.

Analyzing the Relationship between Variables

Today's fire protection managers want to do more than describe a variable; they will want to show how one variable affects another. For example, the chief of personnel for a large fire department may be concerned that the women employees seemed to be promoted more slowly than men. Or maybe a fire departments fire prevention officer suspects that African American citizens are more susceptible to fire losses than white citizens. A newspaper story may have quoted African-American community leaders commenting that local fire officials are prejudiced toward the African-American community. A chief fire Ross said may suspect that a dumpster is responsible for widespread illness in the neighborhoods close by.

Examining the link between such variables is a necessary step in fashioning solutions to deal with these kinds of problems. If men are promoted faster than women are, personnel policies can be modified to ensure that qualified women are part of the promotional pool. If African-Americans are more susceptible to fire losses than whites, fire prevention officers can be trained to build support and trust and have their guidance excepted in a minority community. If illness is more likely to occur in neighborhoods bordering a dumpster compared to those farther away, the site can be cleaned up or citizens relocated. Moreover, each one of these potential links raises an implicit research question that can be stated as a question or what is called a **hypothesis**. One might ask, for example, are men promoted faster than women or are African Americans more susceptible to fire losses than whites? The same questions can be formulated as hypotheses: men are promoted faster than women and African Americans are more susceptible to fire losses than whites.

How do we answer such questions or tests such hypotheses? The answer is by examining the relationship between two variables. One way to do this is by constructing a table, called a **contingency or cross-classification table**, where the values of one variable are cross-classified with the values of another. To understand how this is done we must recall from our statistics courses the discussion of frequency distribution. A **frequency distribution** displays the values of one variable and the frequency with which each value occurs. One might, for example, encounter the following frequency distribution regarding promotions in an agency of city government.

Table 2-1 shows that among the employees in fire and emergency services in 1999, 22 were promoted during the first year of employment, 26 their second year, 29 their third year, and eight their fourth year. The letter "N" equals the total number of employees. These numbers can be converted, and often are, to percentages and yield the following percentage distribution. This table shows that 26% of the workforce was promoted their first year of employment, 31% their second and so on. These numbers are obtained as described by dividing each frequency by the total in the distribution (85) and multiplying by a hundred. For example, 26% equals 22 divided by 85 multiplied by a hundred. Although there is only one distribution in table 2-2, the use of percentages rather than frequencies facilitates comparison among distributions and makes it easier to draw conclusions from them.

Suppose a complaint has been filed by an employee in fire and emergency services alleging that men are promoted more quickly than women are. The charge calls for a comparison of how quickly men and women are promoted. It is necessary to compare the distribution of men with the distribution of women. Another way to say this is that one must cross classify year of promotion with gender. Table 2-3 does this.

Table 2-1: Year of promotion figures for the City of Gothum Fire & Emergency Services year 1999.

Year of Promotion	City of Gothum
First year of employment	22
Second year of employment	26
Third year of employment	29
Fourth year of employment	8
N	85

Table 2-2: Year of promotion figures for the City of Gothum Fire & Emergency Services year 1999.

The first year of employment	26%
Second year of employment	31%
Third year of employment	34%
Fourth year of employment	9%
Total	100%
N	85

Table 2-3: Year of promotion figures for the City of Gothum Fire & Emergency Services year 1999.

	Men	Women	Total
First year of employ-ment	14	8	22
Second year of employment	12	14	26
Third year of employment	7	22	29
Fourth year of employment	4	4	8
N	37	48	85

Table 2-3: Year of promotion figures for the City of Gothum Fire & Emergency Services year 1999.

Year of promotion	Men	Women	Total
First year of employ-ment	38%	17%	26%
Second year of employment	32	29	31
Third year of employment	19	46	34
Fourth year of employment	11	8	9

Table 2-3: Year of promotion figures for the City of Gothum Fire & Emergency Services year 1999. (Continued)

Total	100%	100%	100%
N	37	48	85

Based on table 2-3, is the complaint justified? Because there are fewer men than women, it is difficult to draw a conclusion based on frequencies. It would be easy to draw a conclusion if the number of men and women were equal. As they are not, it is necessary to transform or scaled the frequency distribution so that numbers are based on the same standard. This is done by converting the frequency distributions to percentage distributions, in effect imposing the same scale on the frequency distributions for men and women and facilitating comparisons between them. This is done in table 2-4.

A quick glance at the columns headed men and women reveals that men are promoted earlier in their careers than women. Whereas 38% of the men are promoted their first year, only 17% of the women are. 32% of the men are promoted their second year, compared to 29% of the women. Note that the column at the right is the percentage distribution for the year of promotion for men and women combined shown earlier in table 2-2. This distribution is referred to as the marginal distribution for year of promotion.

In assessing the relationship between two variables like the one shown in table 2-4, a human behavior analyst looks for differences between or among categories of the independent variable. Recalling that the independent variable is hypothesized to influence or act upon the dependent variable. In table 2-4, the independent variable is gender. The dependent variable is the year of promotion. With us, in assessing the relationship between gender and promotion, compare men to women. When you are the analyst, the independent variable will be readily apparent. You, after all, will formulate the hypothesis and should have a clear idea as to which is the **independent variable** or "cause" and which is the **dependent variable** or "effect" variable. When this happens, use common sense. Consider table 2-4. Promotion could not influence gender, so gender must be the independent variable. Assuming or establishing a temporal sequence between the variables may also help in determining which the independent is and which the dependent variable is. Again, with respect to table 2-4, it is clear that gender occurred before promotion. In some situations, however, there are no clues. Consider, for example, the relationship between the number of fire victims over a period of time and the number of firefighters employed by cities of a certain size. Which is the independent variable? Common sense might suggest that it is the number of firefighters that influences the number of fire victims: more firefighters would, one hopes, lead to fewer fire victims. However, cities with larger numbers of fire victims may be pressed into employing more firefighters, so that more fire victims produced more firefighters. In such cases, sorting out which is the independent and which is the dependent variable may be impossible, especially in non-experimental research.

Cost-Benefit Analysis

Herald as a move that would save the taxpayers millions of dollars **"cost benefit analysis"** first debuted in the Department of Defense in 1961 by Rob McNamara. Its purpose was to save the taxpayers money as well as lead to efficiency in government and make decisions more scientific. However, in addition to its failure when applied to the Vietnam War by McNamara and other Defense Department Personnel, its usefulness has been challenged in other settings. Of significance, the technique has been widely used by the Army Corps of Engineers, yet the Corps usually fine dam building projects to be of net benefit, while environmental groups using the same techniques find a net cost. Utility companies display costs-benefit analysis showing the benefits of building a nuclear power plant at a give insight, while local organizations use the same techniques to show that the course are far higher and benefits much lower than claim by utility companies. Costs-benefit analysis has been found useful in a wide variety of settings, yet it may be difficult to except the credibility of a technique that seems to give each interested party to answer it wants. Nevertheless, cost-benefit analysis is widely used and here-to-stay and it behooves Fire Protection Managers to know it. For the remainder of this chapter we will examine the technique and point out its strengths and limitations. Additionally, several how to do its steps will be described.

Suppose you are a planning officer for a fire department in a city of 100,000 people. Your fire chief indicates that a new fire station is needed in the northwest side of the city where a substantial population growth has taken place. He believes this would enable the department to reduce the time per call in that area and dust decrease the cost of fire damage to property owners as well as reducing the risk of injury and death from fire. The fire chief asks you to make a systematic investigation of the costs and benefits of several alternatives, continue without the new station or build one at one of four possible sites. Following your investigation you want to make a recommendation to your department on the relative utility of each course of action. Subsequently, your analysis and recommendations will be the basis of the department's recommendation to the city in its annual budget and capital construction requests.

The procedure you are asked to undertake is called the cost-benefit analysis. Its primary rationale is that when choosing among alternative courses of action you should pursue that which produces the greatest net benefit. Net benefit, of course, is the total benefit minus the total cost. Ultimately, this rule is the fundamental rule of cost-benefit analysis.

What is the advantage of cost-benefit analysis? Its biggest contribution may be that it forces the analysis to think about the actual costs and benefits of alternative choices. In doing so, course and benefits that have not been part of the decision may come into light. Or one alternative may be found to have such high course that it can be rejected out of hand. Cost-benefit analysis forces you to make explicit the foreseeable costs and benefits. Ultimately, any decision is based on some sort of calculation of course and benefits. Sometimes the calculation is on unconscious. Other times it is explicit but not very well thought out. In consciously applying cost-benefit techniques the aim is to bring to light all the course and benefits that can be foreseen and try to assess the relative weights.

What are some limitations of cost-benefit analysis? Cost-benefit analysis has certain simplicity to its ring. It sounds scientific, it tries to compare apples and apples and not apples and oranges. Just how do you calculate and translate cost and benefits into dollars? When you're choosing whether or not to buy a new car, certain course and benefits are relatively easy to calculate. Cost seems easy to estimate, the price of the car, interest to finance it, cost of gas and repairs over the years, and license fees, for example. Some benefits are more difficult. You might be able to estimate the costs of not having a car, having to take public transportation, for example, and included as a benefit for each of the alternatives. But how do you estimate the value of convenience in owning a car beyond the estimate of course of alternative transportation? And in choosing between the alternatives how do you assess the dollar value of your preference, let us say for a new car over an old one or a sports car over a minivan? Obviously you can make estimates of the dollar value of such things as convenience and accessibility. However, it is easy to see that even in this simple example, values quickly enter into the assessment of course and benefits. And, of course, the values placed on various kinds of course and benefits will determine which alternative has the law just net benefit. So we can begin to see why cost-benefit analysis done on the same project by groups with different basic values might come to entirely different conclusions.

Another limitation is that some costs-benefit analysis, only consider one's own costs and benefits, and not the costs and benefits a decision may impose on others. These are called **externalities.** In the case of a decision whether or not to buy a new car, the cost-benefit analysis indicated that keeping the old jalopy maximized net benefits. However, it may be that the old jalopy is an environmental disaster, creating much more air and noise pollution than a newer car. Subsequently, others pay the cost to, in terms of health as well as clean up, for your old jalopy.

In pursuing his directives to make recommendations for a fire station based on cost-benefit and other appropriate criteria, the fire department planner has collected the information in the following table.

Table 2-4 Costs and Benefits of Alternative Sites

Station site	Construction	Operation	Land acquisition and relocation	Benefits	Net benefits
1	3	4	6	15	2
2	4	4	3	17	6
3	3	4	10	20	3
4	5	4	4	18	5

The fire department planner figures the benefits as being property saved from fire destruction by the location of a fire station at each particular site. This figure could depend on the value of property near the site, the distance of the fire equipment from other sites, which would allow some measure of response time, and the fire proneness of each area, based on past experience and age and condition of the buildings. The plan also did similar calculations for the value of human life saved.

The planner calculated that site too had the greatest net benefit of 6 million. Site for what they net benefit of 5 million was second. Before recommending site to the planet made the following observations. Neither site to your site for had a disproportionately negative impact on various ethnic or income groups in the area, the community resistance to locating a fire station at either site to or for was minimal since neither was in the immediate for sanity of a residential area and there would be no exceptionally negative environmental impacts at either location. The planner is therefore recommended site to as being the best bargain for the city.

You have gleaned over an overview of the problems and possibilities of cost-benefit analysis. Remember that while cost-benefit analysis can be a useful tool, it is not the only way of examining a problem. Remember, too, that the seemingly hard data of cost-benefit analysis are based on very slippery estimates of the value of everything from human lives to interest rates. In examining and evaluating cost-benefit analysis done by others, you must discover and evaluate the value judgments and went into the estimates. What was undervalued? Overvalued? Or omitted? Likewise when you are preparing your own cost-benefit analysis, be explicit about values assigned to various factors. Used with care and common sense, cost-benefit analysis can be a useful tool in "Fire Related Human Behavior." It does help make explicit the kinds of course and benefits occurring in each option under consideration. Whether it is worth the time and trouble can only be determined in each individual case. And use carelessly, cost-benefit analysis, like any other technique, can be highly misleading and even fraudulent. Because the final results look so firm, it is incumbent on the Human Behavioral Scientist to exercise special care in using this form of analysis.

Chapter Two Summary

Look over the following scientific claims and check whether they are true or false according to your knowledge or whether you are not sure about them based on current available data.

1.	Smoking marijuana causes damage to fetuses	True	False	Not sure
2.	Marijuana is not addictive	True	False	Not sure

The issue is not that scientific claims are obsolete true or false but rather that claim's are really hypotheses that have received varying degrees of support from the available scientific research. Thus, the hypothesis that smoking marijuana damages fetuses has strong support, and probably is true. On the other hand, the hypothesis that marijuana is not addictive is false. However, in most cases, there is not much scientific evidence one way or the other, either because weaknesses in the research design, which is often unavoidable, make the evidence inconclusive or because few researchers have yet conducted research on the topic.

Read the following research report and write a critique of the experiment based on what you learned in this chapter. Address questions such as what type of research does the report exemplify? What are the dependent and independent variables? What control methods were used and how reliable are the data?

Fire Company Staffing Requirements: An Analytical Approach
By
John Jay College Adjunct Professor Peter Blaich

Introduction

In today's world, the risk of danger has increased tenfold. The existence of a skilled firefighter staff in each and every community is of critical importance, as these individuals served as the first line of defense in response to emergency situations. Although the events of September 11 increased the importance of firefighters in communities, economic downturns have required that many community fire departments cut their firefighter staff to bare minimum levels, and this circumstance results in tremendous impacts for retained firefighters as well as community residents. The minimum standard for him and his simple fire department staff is a fiercely argued topic, as the standard varies widely from one community to the next. The number of firefighters permitted within a given department can ultimately influence the degree of stress and injury that is encountered within emergency settings.

Summary of Literature Review

A wide body of literature exists regarding this topic and the consequences for firefighters in the line of duty. In 2003, Sherwood L. Boehlert, chairman of the committee on science in the United States Congress drafted legislation that includes the following needs assessment. An estimated 73,000 paid firefighters served in departments protecting 50,000 or more community members that have few event for firefighters assigned to an engine, which is the minimum national standard (Boehlert 8). This legislation is poised to directly combat this problem in that it will assist fire departments across the nation in meeting new minimum staffing standards in order to effectively manage fires in acts of terrorism (Boehlert 8).

Firefighter fatalities in recent years have been attributed to a number of circumstances, including heart attacks, dramatic injuries from a vehicle in training accidents, and other related situations; however, the number of fatalities is on the decline, while the number of injuries is on the rise (United States Fire Administration 8). Repetitive dramatic workplace event exposure may result in increased levels of posttraumatic stress for many firefighters, and this may be a consequence of the requirement to serve in the large numbers of emergencies over time (Beaton et. .al 822). The September 11 attacks in New York City demonstrated that the need for safety controls that must be in place for all rescue personnel, regardless of the hazard (Spadafora 536). Furthermore, during prolonged operations, it is advised that the number of hours for working firefighters is limited during prolonged operations, such as 12 hour shifts, in order to reduce physical and psychological stress over the long term (Spadafora 537). Finally, it is encouraged that firefighters gain access to regular counseling on a mandatory basis, so that personnel can effectively manage the stress and outcomes of their daily work activities (Spadafora 537).

Legislative officials in the United States have attempted for years to recognize the importance of staffing concerns by municipal fire departments. According to data provided by the U.S. Fire Administration, approximately 100 firefighters died within an average year, and many more are injured, and many of these injuries are directly related to inadequate staffing (Boehlert 1). For example, it has been demonstrated that in communities where crew sizes were comprised of less than four firefighters, 69% had injury rates of 10 or more per hundred firefighters, and only 38% of areas with crew sizes of four or more members had similar injury rates (Boehlert 1). In communities where firefighter staffing levels are at the bare minimum, it is anticipated that the loss of lives of firefighters and community members will be much greater, as fewer personnel are available to handle emergencies (Boehlert 1).

The frequency of workplace injuries is directly correlated to gender, each, and tenure (Liao et. al 229). Prior research indicates that many male firefighters will not disclose minor injuries in the workplace for a variety of reasons; however, female firefighters are the likely sufferers of increased numbers of injuries (Liao et. al 239). Since workplace injuries are coarsely, the importance of identifying the reasons for the increased likelihood of injuries is critical, and in the case of firefighters, inadequate staffing levels may be to blame (Liao et. al 240).

In 2000, the National Fire Protection Association reported 84,550 firefighter injuries that required medical treatment or that required at lease one day of restricted activity, and the leading cause of these injuries was over exertion (Walton et. al 454). Prior research indicates that the injuries that are caused by over exertion can result in significant problems for fire departments, and the primary contributors to such injuries include heavy lifting; staffing, training; unsafe posture, inattention and fatigue, and other physical or environmental conditions (Walton et. al 457). By eliminating these injuries, it is estimated that $319 in direct cost would be saved per firefighter on an annual basis; therefore, it is critical that additional emphasis is placed on the degree to which these injuries are caused by factors directly related to staffing concerns (Walton et. al 458).

The firefighting profession is a physical and emotional demanding set of circumstances, whereby saving lives is of primary importance. It is common knowledge that a large amount of the typical fire department budget is allocated to personnel costs, with as much as 90% required for salaries (Lawrence 202). The necessity for increased staffing can result in a tremendous burden for fire departments if their budgets are reduced or maintained at an appropriate level (Lawrence 202). It is also known that the nature of firefighting requires prolonged exposure to substantial risks, including altered metabolism, decreased alertness, increased fatigue, and sleep deprivation (Glazner 262). Each of these factors contributes to the increased risk of injury for firefighters, particularly for those working long hours due to inadequate staffing levels and municipal departments (Glazner 262).

Fire department under staffing can potentially result in the dues to fire service effectiveness, increased property losses, and reduced life safety factors for firefighters and ordinary citizens, and this demonstrates a critical disregard for the traditional functions of a municipality as a political entity in charge of its residents (Lawrence 210). In general, a strong correlation exists between the number of firefighters on duty and available for emergencies and the number of injuries that are sustained, as well as the number of fires within a given period of time and the number of injuries that occur as a result of these emergencies (Vatter 127). If staffing levels are restored to sufficient levels within a municipality, then it is critical that the severity of fires is considered, since this can directly impact the number of injuries that occur (Vatter 127). Finally, it is a general rule that the number of firefighters that respond to a fire is directly related to injury statistics, and this can be influenced dramatically by staffing shortages and cuts (Vatter 128).

Methodology

The following study will attempt to identify the linkage between the staffing levels of a fire department and the increased risk of injury in the line of duty related to performance issues due to inadequate levels of staffing. Specifically, it is assumed that reducing staffing numbers with in a fire department will lead to increased risk of injury while participating in emergency situations.

Hypothesis: The small of a firefight up through the greater the chance of firefighter injury.

Unit of Analysis

The unit of analysis that will be used for the proposed study is the firefighter with in an average municipal fire department. The term "average" will be described in detail to determine the appropriate variables required for inclusion in the study.

Dependent and independent variables

The primary dependent variable in the proposed study is that proper and success-ful performance is directly identified with how many firefighters properly per-form firefighter tasks in the line of duty, and this concept is enjoyed by the successful completion of firefighter tasks in the line of duty. The primary inde-pendent variable is firefighter crew size, which will be identified by a number of minimum standards, including the requirement for 3, 4, or 5 person teams.

Control variables

The Four primary control variables for the proposed study are the following: 1) the injury of a firefighter, which is the rate of injury compared between operating crews of 3, 4, or 5 person teams, we'll be identified with information regarding the minimum standards for various municipalities; 2) The level of physical fatigue, which is the exhaustion level that a firefighting team demonstrates in comparison to different crew sizes, and will be made available with municipality data; 3) The time that is required to complete tasks is the amount of time that each crew size needs to respond, gain entry, ventilate and raise ladders, apply water through hose lines, and search and rescue; and 4) The time that is required to rescue victims is the minimum time that is necessary to rescue a civilian, which will be acquired from various municipalities.

Sample selection

The study population includes firefighters that operate in various municipalities, and the target populations which include fire Department crew teams that oper-ate under a 3, 4, or 5 person crew in areas with populations of 150,000 or more.

Data collection

Data will be collected through stratified sampling, where the strata are levels of operating crews with 3, 4, or 5 persons. As the size of the operating crew is determined, data will be collected from municipalities with a variety of staffing levels, and these firefighters will serve as the sample population for the study.

Statistical Analysis

Once the data are collected, we must construct a portable casual model predicting firefighter injury as a result of staffing reductions. Assuming the dependent variable is an interval scale. The casual analysis of equation will be: $Y=a+b1x1+b2x2+b3x3+e$

This research has given light to a plausible argument that some independent variables are caused by other independent variables. By hypothesizing that the size of the firefighter crew has a direct effect on firefighter injury with that crew, but that proper and effective performance of task has an indirect effect on firefighter injury. The variables used to predict injury include:

$X1$= independent variable= the size of the firefighter crew
$X2$= control variable 1 = rate of injury to firefighter
$X3$= control variable 2 = physical condition after completion of the task
$X4$= control variable 3= time that is required to complete tasks
$X5$= control variable 4= time that is required to rescue victims

The primary sampling units (PSUs) for the proposed study will be a group of municipalities that in comp is a further subdivision of strata that are classified by a 3, 4, or 5 person team, and the sample location will be identified according to all available data that will be collected from various municipalities that currently operate at the staffing levels.

Other Issues

The proposed study is likely to result in other issues of concern, which may include current and future budget allocations for firefighter personnel, prop of firefighter training, and other creative staffing decisions. The data that will be collected and analyzed in this study is likely to result in a strong consideration of the enormous staffing problems that are currently experienced in various municipalities across the nation.

Data Matrix

Unit of direction	Independent variable	Dependent variable	Control variable number one	Control variable number two	Control variable number three	Control variable number four
Fire-fighter crew	Staffing of 3, 4, or 5 person crew	Proper and effective performance of tasks	Rate of injury to firefighter	Firefighter physical condition after the completion of tasks	Time that is required to complete tasks	Time that is required to rescue victims
1	Three-person crew		7.77 injuries per 100 firefighters	Complete exhaustion	18.8 minutes	
2	Four person crew		5.3 injuries per 100 firefighters	Near exhaustion	10.29 minutes	80% faster
3	Five person crew		5.3 injuries per 100 firefighters	Little evidence of exhaustion	6.15 minutes	

References

Beaton, R., Murphy, S., Johnson, C., Pike, K., and Corneil, W. "Expose to Duty Related Incident Stressor in Urban Firefighters and Paramedics." Journal of Traumatic Stress, 11.4 (1998): 821-828.

Glazner, L. "Factors Related to Injury of Shift Working Firefighters in the United States." Safety Science, 21 (1996): 255-263.

Lawrence, C. "Fire Company Staffing Requirements: An Analytical Approach." Fire Technology, 37 (2001): 199-218.

Liao, H., Arvey, R., Butler, R., and Nutting, S. "Correlates of Work Injury and Duration among Firefighters." Journal of Occupational Health Psychology, 6.3 (2001): 299-242.

Spadafora, R. "Firefighter Safety and Health Issues at the World Trade Center Site." American Journal of Industrial Medicine, 42 (2002): 532-538.

United States Fire Administration. "Firefighter Autopsy Protocol." (1994): 1-26.

Vatter, M. "The Impact of Staffing Levels and Fire Severity on Injuries." Fire Engineering, 152.8 (1999): 125-129.

Walton, S., Conrad, K. Furner, S., and Samo, D. "Cause, Type, and Workers Compensation Costs of Injury to Firefighters." American Journal of Industrial Medicine, 43 (2003): 454-458.

Discussion and Review

1. Can you identify the persuasive techniques used in each of the following statements? Recent and dramatic research has indicated that high school students who participate in after school activities have a higher rate of completing high school than similar high school students who did not participate in after school activities. Although most criminologists believe that incarceration is an ineffective method of crime prevention, every person knows just how effective incarceration can be when used consistently. Research has shown that illegal aliens would become citizens if they were able to do so. However, this research was conducted by a rich elitist psychologist who knows a little about the plight of immigrants.

2. Presently is the "Scientific Behavioral Method" popularity and has it been effective in improving our standards of living?

3. The book interprets the term "significance," explain the three different types associated with this scientific term and to which one human behavior researchers rely on mostly.

4. List and describe the most widely used persuasive techniques as explained in this textbook.

5. Define the term "dependant" and "independent variable."

6. What are the findings of Latane and Darley's Experiment?

7. Name one of the most widely used hypothetical constructs?

8. Why do most behavioral researchers indicate whether their findings are statistically significant or not?

9. Most research is reported to have results between what levels of significance?

10. Why do you want to know how many times the researcher has replicated the experiment?

11. Can you name the method that scientists and behavioral researcher's alike use to improve confidence in their research?

12. Why do you think that most case studies concerning Fire Related Human Behavior use the Correlational Method instead of the Experimental Method?

13. What does a result of zero (0) indicate?

14. What is the minimum group size preferred in gathering statistics for correlational analysis?

15. Can you explain the term reliability as it refers to research data?

16. Can you explain the consistency issue referred to as internal reliability?

17. Is there an internal reliability problem associated with the National Fire Incident Reporting System (NIFRS) data system?

18. Validity is questionable to what extent?

19. Can you explain the term external validity?

20. What is the most common method of sampling?

21. What is the value of the correlational method?

22. What must remain the core competency or functional area of expertise for the Nation's Fire Service?

23. A question in research is known as the?

24. Can you explain what a contingency or cross-classification table is?

25. What is frequency distribution?

26. What does a human behavior analyst looks for?

27. Name the scientific term for "cause" and "effect?"

28. What are some limitations of cost-benefit analysis?

29. What is the advantage of cost-benefit analysis?

30. Identify and explain the difference between the two types of research techniques that you may encounter?

3

A Systems Approach To Fire Related Human Behavior

It is difficult to predict, and impossible to understand, the behaviors of people without some knowledge of the goals that they are pursuing.

—*Professor Norman Groner*
John Jay College of Criminal Justice

Introduction

Is it possible to separate the performance of buildings, hardware and people? The answer is no! In fire emergencies, the performance of buildings and people are highly interdependent. How will people perform during a fire emergency depends heavily on the building. Of significance, the converse also is true. The performance of a building in detecting and limiting the growth and spread of fires often depends heavily on the actions taken by people in the structure. Moreover, buildings determine the actions that people take by providing the information they receive during an emergency. Some of the ways in which the performance of people depends on buildings and their fire protection features include some of the following.

The speed in which a fire can be detected and whether its location is accurately identified depend on the layout of the building, its detection systems, especially alarmed system annunciator displays and vocal alarm systems, and how various areas are used. For example, fires in concealed spaces are notoriously difficult to locate and vocal systems for reporting the exact location of a fire at proven very effective.

The buildings fire alarm signal can notify people that there is a potentially life-threatening emergency, or people can interpret it as just another nuisance or false alarm, a fire drill, and alarm test, or even another type of emergency altogether. Of significance, people exposed to frequent false alarms are much less likely to respond to fire alarm signals.

An alarm signal may provide people with useful information about how they should respond, or it may add to their confusion and anxiety. For example, if a favorable fire alarm transmits an unintelligible message; this is likely to increase the anxiety of building occupants.

The signs in a building may help or hinder efforts to find egress routes. For example, there is evidence that posted floor plans actually can slow people in finding their way. Signs also can mislead or confuse. Moreover, a sign with a message that reads "use only in the event of an emergency" may lead people to wait at the exit unless they are certain that there really is a fire, even while the alarm is sounding.

The layout and design of a building may help or hinder efforts to locate egress routes and areas of refuge. Mainly, it is more difficult to locate exit doors that look the same as other doors. Irregular building layouts, such as those found in department stores are open office plans, can confuse people, causing them to lose their way while trying to evacuate.

Building performance during a fire also depends a great deal on "Fire Related Human Behavior." Some examples where people can help or obstruct building performance include the following.

People block open fire rated doors, allowing the unimpeded spread of smoke and fire. Also, people fail to close other types of doors that could slow the spread of smoke and fire.

People use buildings in inappropriate ways that increase the risk of fire. Of significance, people cooking in hotel rooms have started fires. A change in occupancy in an existing building can create unforeseen problems when people use the building in ways the designers did not anticipate. Subsequently, storage of large amounts of combustible materials can overwhelm fire protection systems designed for ordinary hazards.

Additionally, people fail to maintain detection systems, fire sprinkler systems, and other important fire protection features, thus causing these features to fail or malfunction during fire.

Moreover, people forget the locations of fire extinguishers or forget how to use them, thereby failing to extinguish fires during the incipient stage of fire development.

Lists some examples that you have experienced in which the building determined the actions of people, and the actions of people affected the performance of the building.

_____

_____

A Systems Approach That Incorporates Human Behavior

After completing the prescribed exercise, you should understand that buildings, fire protection hardware and people interact in complex ways to determine the outcomes of fires. To understand how people and buildings worked together during fires and emergencies you will find it useful to think of the building and its occupants as part of a single system. People, building features and fire protection hardware are all system components that can and should work together to prevent fire casualties and property losses. Occupants need to take advantage of building features that detect and inhibit fire and smoke. Design and maintenance of buildings should help people choose the actions that prevent ignitions, inhibit the growth and spread of fires and encourage rapid and effective evacuations.

We can divide systems models that usefully incorporate human behaviors into two basic categories, the hard systems approaches and soft systems approaches. Perhaps the best way to define hard systems approaches is to compare them to soft systems approaches. All hard systems approaches differ from soft systems approaches in two important ways.

Hard systems approaches require well defined problems. The user must know all of the important elements in the system and the relationships among them before the model can provide reasonably accurate predictions of systems behavior. Soft systems models are much more open ended. Elements and relationships can be added as the model evolves. Therefore, soft systems models are well-suited to exploratory types of studies where the system is poorly defined.

Hard systems approaches can accept only quantifiable data or numerical input because systems behavior involves mathematical equations. Subsequently, the outputs from the hard systems models are numerical predictions of systems performance. In contrast, soft systems approaches typically allow the expression of qualitative relationships, although they may be able to accept "hard" quantifiable data as well. Moreover, soft systems models typically do not yield precise predictions of systems behavior. Hard systems models are extremely attractive because numerical outputs are specific and seem to be objective. By objective I mean that the results from the model do not depend on the views and options of the persons using the model. However, this appearance of specificity and objectivity is often more an illusion than real. The truth is, the results of hard systems models often is too sensitive to predict assumptions built into the model and users can hide or overlook these local these assumptions easily. For example, a quantitative model of fire growth in a hard systems model is extremely sensitive to such seemingly minor factors as ventilation. Thus, whether a window is closed or open just a little can radically change the way a model predicts fire growth. However, to someone unfamiliar with how the model is constructed; the results may seem to be accurate without regard to ventilation. Another example involves the modeling of evacuation times. A naïve user could easily take these times as objective and specific. The truth is many models based their outputs on optimal of evacuation times, which do not consider all of the important factors such as delays due to the ambiguity of clues and preparations before of evacuation that influence of actuation times in the real world. We can adjust such results with safety factors are aiding time to take into consideration the fact is omitted from the models, but again, the values used for the safety factors are subjective. The omission of factors that are not easy to include in hard systems models leads us to the second shortcoming of hard systems models. Just as important as the misleading appearance of specificity and objectivity is the ability of hard systems models to handle only those types of input data that are consistent with the model. For example, a faulty tree another hard systems model can use only data about types of faults and their probabilities of occurrence. These models typically only at faults were probabilities are unavailable; this is often the case for human errors. Also, faulty trees cannot handle data about people's abilities to troubleshoot and correct unforeseen faults. These types of data do not fit the model and must be omitted from the calculations. Like faulty trees, most hard systems approaches do not easily incorporate human behaviors that involve people's abilities to make decisions and solve problems. For example, hard systems models are ill-suited to calculating whether and when people decide to evacuate upon hearing a fire alarm. This problem is difficult to quantify because many factors interact in complex ways to determine when people decide to start with their evacuation. Subsequently, neither of the computer models discussed in this chapter directly models this decision-making behavior, even though its importance to system performance is apparent. However, they do compensate in part for such missing elements by using behavioral rules for example parents always attempt to rescue their children and safety factors for example most evacuation travel times are doubled or by adding delays for example assuming building occupants will start to evacuate 10 minutes later after hearing a fire alarm signal.

A useful alternative to Hard Systems is Soft Systems Models. **Soft systems models** lack the specificity of hard systems models. Ultimately, they have the considerable advantage of being able to incorporate a much greater range of systems elements and relationships among those elements. There are many types of soft systems models, but few researchers have applied them to fire safety. An exception is goal-based models, which will be discussed later in this book.

To date, Behaviorists have used hard systems approaches principally to model evacuation times from buildings. Models used for this purpose fall into the following two basic categories.

Standalone egress width or evacuation time models, help calculate required egress widths and capabilities and optimal evacuation times from buildings. Most of these methods are simple enough that they do not require computer programs.

Egress software modules that are part of larger scale computer programs used to model many aspects of fire incidents. These modules interact with other modules, especially those that predict the growth and spread of fire and the products of combustion. Egress software modules depend on some of the methods described in the first category, but they generally are used in conjunction with fire growth software modules to try to gain a relatively complete picture of how a fire incident can be expected to develop. Typically, fire growth of modules help calculate how much time is available to occupants before conditions become untenable. Egress modules calculate the time needed for people to escape before conditions become untenable. To computer programs **HAZARD I and CRISP** include well-developed egress software modules that predict human behaviors.

Hard systems approaches have found wide acceptance from the **Society of Fire Protection Engineers** as a means to calculate exit widths and evacuation times. These methods depend, in varying degrees, on the subjective judgments of code writers and on calculations derived from empirical studies of evacuation of buildings.

A Fresh Look at Fire Protection Engineering

Among the earliest requirements incorporated into the building and fire safety codes were the design of exits and buildings. These requirements have evolved over time and rely on the judgments of code writers; therefore, some have criticized these codes for the lack of basis in research. Nonetheless, they have served good over many decades, despite their technical inadequacies. These codes are similar to hard systems approaches because they yield specific quantified results taste on numerical input. However, it is worth noting that the system in question typically includes only a few of the components that determine how well people function during a fire or emergency. Also, they differ in that hard systems approaches incorporate clear rationales rather than unexplained judgments.

Until recently, the **Lane Model** has been the basis of code provisions that govern the width of exit stairs. Of significance, this model stems from the way cars travel on roads; a road with two lanes in each direction is assumed to carry twice the number of cars as a road with one lane in each direction. For stairs, a lane is about 22 inches wide. This number is based roughly on the shoulder widths of men. Subsequently, on a 44 inch stairway, we assume people position themselves shoulder to shoulder as they descend. Using the Lane model, we conclude that stairs 40 inches wide should carry roughly the same number of people as stairs that are only 22 inches wide. But we assume 45 inch wide stairs will carry twice the number of people as 40 inch wide stairs. The flow rate traditionally used for calculating evacuation is 44 persons per minute per 22 inches of stare width. Experts derived this figure from a French study where firefighters were asked to deliberately hurry to achieve the greatest possible rate.[17] Of significance, I have found and you will learn that this model makes faulty assumptions about how people really move down corridors and stairs.

Building codes and the National Fire Protection Association (NFPA) Standard 101, *Life Safety Code*, uses a straightforward approach for calculating minimum widths for corridors and stairs. "The required width, in inches, of the exit shall not be less than that obtained via multiplying the total occupants load served by an exit by 0.34 stairways and 0.24 other exits are less than the minimum widths specified elsewhere in this code."[18]

Later in this textbook we will discuss the model codes for details about how to perform the calculations. In this chapter we are more concerned with the assumptions about human behavior that underlined these calculations than with the mechanics of performing them. Of significance, the underlining assumptions of these calculations are a mix of empirical evidence and judgment. Research by Paul's in 1987 focused on actual evacuations demonstrated that the Lane model did not accurately predict the rate at which people evacuated. Observations of people using stairs revealed that the Lane model has two fundamental problems. First, it does not accurately represent the way that people actually descend stairs. People sway back and forth and try not to touch walls or each other. Therefore, they take up considerably more room than 22 inches and tend to stagger themselves rather than travel side-by-side. For this reason, the flow rates used in the traditional models were too unrealistic. Paul's also discovered that flow rates do not increase in a stepwise fashion. Instead, flow rates are linearly related to stare width; beyond a certain minimum width, even a small increases in stair width increased flow rates. Moreover, Paul's demonstrated that because of an edge effect, we should subtract that out or 6 inches on both sides of stairs from the measurement of stair width when we calculate flow rates. This edge effect results from the swaying motion that people make while moving down stairs, and from their desire not to touch walls and each other. Because of this research, code writers abandon the Lane approach. Of significance, the calculations described above use a linear model.

A considerable amount of research exists concerning how quickly people can evacuate from buildings using stairs. Stairs are a critical method for evacuating occupants from all types of buildings. In certain buildings, most notably large office buildings and places of assembly, the carrying capacity of stairs directly correlates with the time needed to implement a life safety strategy. However, the capacity of stairs is only one of several methods that we need to consider when planning for emergencies in high-rise buildings and places of public assembly. Moreover, Paul's has pointed out, the time needed to detect a fire, alert occupants, and then organize and begin and evacuation can equal or exceed the time needed for occupants to actually travel to the outside of the building. In 1969, the National Research Council of Canada initiated extensive research and building egress problems. Between 1969 and 1974, Paul's and his colleagues observed and collected detailed data during nearly 30 of evacuation drills, primarily in high-rise office buildings. They used participant observers and video and audio tape recordings to compile extensive quantitative and qualitative information about flow rates in stairways and the factors that affect those flow rates. Most importantly, Paul's was able to use though collected data to build good statistical models of evacuation flows and total evacuation times. These models yielded predictions that differed dramatically from the flow rates upon which code writers traditionally had based code provisions.[19]

Paul's explains the flow in which crowds move downstairs and corridors is most simply expressed as a linear function of the following three measures.

FLOW=SPEED x DENSITY x WIDTH

Where:

Speed = the distance covered by a person over a certain unit of time, for example, 3 feet per second or 180 feet per minute;

Density = the number of people occupying a certain space, for example, 2 persons per square yard, or, conversely, 0.5 square yards per person; and

Width = the width of the stairway or corridor.

Remember to keep in mind that the width used in the calculations should be somewhat less than the actual measured width because people sway when a walk and try to avoid brushing against the edges of stairs and corridors.

According to the model, as speed, density, or width increases, the flow of people increases proportionately. This means if the density is doubled then the flow rates also should double. Moreover, the reality of the situation is much more complex. The affects of speed and density on flow rates are not independent; as density increases beyond a certain point, people begin to interfere with each other's ability to maintain a normal gait, and they must slow down. Plus, when density increases beyond a certain point, the flow rate begins to level off and even slow somewhat. Based on observations by Paul's the best possible flow rate for the sending stairs is about 1.18 persons per second per meter of effective stairway width. This rate is ultimately achieved in no more than one half of the uncontrolled, total evacuation was in office buildings in which such evacuation procedures are well known to all occupants.

Another important finding from Paul's research is that the actual carrying capacity of stairs is considerably less than the carrying capacity the traditional Lane model provides. Compared to the assumed 44 person per minute on a 22 inch wide stair, the actual flow was, at best, about 24 persons per minute.

The considerable inaccuracy of the traditional models raises a question; if the traditional Lane model so obviously over estimates the flow rate for stairs, why haven't we had tragedies where occupants put in evacuation because the stairs were already full of people? Of significance, high-rise buildings have an excellent record for very low rates of fire casualties. It is my belief that part of the answer lies in examining the very conservative assumptions that code officials make when they calculate the number of people likely to be in a building. The actual number of persons in a high-rise office building is often less than half the number codes assume. Moreover, not all building occupants start evacuation at the same time! Clearly, the error is made in calculating the number of occupants and the speed at which they can evacuate counteract each other, thus reducing especially congested hazard as situations in high-rise buildings. Of significance, it is important to note that many high-rise occupants will have to queue before entering stair-ways when they start their evacuation at the same time. Since this can happen on a floor affected by fire, it is often inadvisable to use a life safety strategy based on a completely uncontrolled evacuation of all building occupants.

Once we know the flow rate for exits in the total population of a building, it would seem easy to calculate the total time it would take to evacuate the building. Moreover, if everybody started toward the stairs the moment they heard a fire alarm sound, you would probably have an accurate estimate of the total evacua-tion time. However, people do not behave like that and there is a start up time before exit paths reached their full capacity. Again, Paul's, turned to data from actual building evacuation is to find out how quickly people really evacuate buildings. He found an average startup time of about 41 seconds for the build-ings his team observed. By combining data from start up times and the effects of population on flow rates down stairs, Paul's developed the following equation for predicting total evacuation time from high-rise buildings with fewer than 800 occupants.

$$T = 2.00 + 0.0117p,$$
Where:

T = minimum time, in minutes, to complete an uncontrolled total evacuation by stairs; and

P = actual evacuation population per meter of affected stare with—measured just above the discharge level of the exit.

You need two pieces of data in order to equally estimated optimal total evacu-ation time in high-rise buildings. They are the total population of the building and the number and widths of the exits. No at the widths of the exits should be measured just before where the occupants leave the building, that is, just above the discharge point of the exits.

Let's walk through the following evacuation calculation together: There are 700 occupants in a building with three exits. Two of the exits are 40 inches wide and the other exit is 60 inches wide.

First, convert the actual width of each exit to a measure of affective with by subtracting 12 inches from the width of each exit. This will account for the edge effect of each of the two sides. Thus, the effective widths of 2 exits become 28 inches, and the affective width of the remaining exit becomes 48 inches. The total affective width of the exit is the sum of the affective widths, that is, 104 inches.

The equation is expressed in metric number instead of inches and feet, so you must convert inches 2 m. 1 inch equals 2.540 cm or 0.0254 meters. 104 inches multiplied by 0.0254 equals about 2.64 meters.

The variable p in the equation is the population size divided by the effected width (in meters) of the exits.

$$P = \frac{Occupants}{Effective\ width} = \frac{700}{2.46} = 265.15$$

With the value of p, you can calculate T, the total time as shown:

$$T = 2.00 + (0.0117 \times 265.15) = 5.10 \text{ minutes}$$

Thus, in our example, an optimistic evacuation time for the prescribed building would be a little more than five minutes.

Assuming that the population in the same building doubles to 1,400 occupants, then, the value of p also doubles and becomes 530.30. Remember that according to Paul's you should use a different formula for populations greater than 800.

Using the recommended formula, the total is calculated as the following:

$$T = 0.70 + (0.0133 \times 530.30) = 7.75 \text{ minutes}$$

Paul's observations revealed to us that several factors can influence evacuation times, including the dimensions of stare treads and risers, and whether building occupants are wearing heavy winter clothing. This was especially notable during Canada winters when Paul's made his observations. Nonetheless, the data indicated that the above formulas provided generally accurate estimates of the total evacuation time in buildings where the occupants responded quickly to the fire alarm. Of significance, you should understand that the equations are intended to predict average evacuation times. They do not predict evacuation times that are worse than average and they especially don't predict evacuation times when things go wrong, as when a fire alarm malfunctions. Where building occupants are unfamiliar with emergency procedures and especially in buildings with false or malicious alarm problems, a start up times certainly will be considerably longer than the times used in the equations. Moreover, Paul's points out, the formula will not yield accurate evacuation times in tall buildings with low populations on each floor. Pauls also cautions that using population figures derived from occupancy load factors in codes could over estimate actual population by twofold or more.

One must also consider that the calculations do not come from realistic fire scenarios. During a real high-rise fire, it is usually important to evacuate the fire and adjacent floors quickly. In an uncontrolled evacuation like those used for data collection, occupants on upper floors may have to wait before gaining access to crowded stairways. Waiting for access to stairs on the fire floor is unacceptable for most life safety strategies. Subsequently, this points to the importance of using controlled sequential egress in life safety strategies for many high-rise buildings. The use of life safety strategies is a soft systems approach that will be discussed later in this textbook. Sequential evacuations require extensive communications capabilities and well-trained building officials and occupants. When these are not present, building evacuation can be prolonged, this organized and during a fire deadly. The effective width model provides a generally accurate approach to calculating evacuation times in high-rise buildings when compared to the traditional approaches the designers of code requirements and buildings have used. Fire Protection Engineer's now have a valid criteria for sizing their exit facilities and the effective width model is now an optional method for calculating exit widths and NFPA 101A, *A Guide on Alternative Approaches to Life Safety.* However, stairs are only one part of what should be an overall strategy for saving lives in the event of a fire or emergency. Additional planning must ensure that the most endangered people have priority access to stairways. Of significance during the evacuation of the World Trade Center, stairs in themselves presented a serious obstacle to evacuation, an issue that will be discussed in a later section.

The computer models discussed in this section are the most sophisticated hard systems models used in Fire Protection Engineering and subsequently incorporate human behavior. Other types of models include event trees mainly fall trees and cause-consequence trees, operations research methods and various types of hazard assessments and risk analysis. Moreover, these other types of models are sometimes applied in such a way as to incorporate certain human behaviors. However, the computer models discussed below represent the most intensive efforts to date that incorporate human behaviors into overall systems models, which is why I have chosen to represent them to you as hard systems models.

HAZARD I is a fire hazard assessment computer program originally developed at the National Institute of Standards and Technology (NIST). It includes various modules that together provide fire protection engineers with an overall view of how fires, building layouts, and people might be expected to interact giving a design fire. The term **design fire** refers to a hypothetical fire with exactly defined characteristics. These characteristics might include such specific information as the source of ignition, fuel, and room geometries. Design flaw is represented fires that I'll most likely to occur in the real world. However, the selection of design fires as the basis of computer models or any other type of simulation does not necessarily predict how a real fire will occur, because these models may fail to consider other factors. Fire protection engineers may omit these other factors either because the model cannot handle the data or because they do not have similar data to use. HAZARD I is intended to model incidence and smaller buildings. It includes a human behavior module called EXITT, along with other modules that predict the growth of a fire. Of significance, a zone model that predicts the height of the interface between the hot upper and cooler lower layers of gas with in the rooms of the building, along with the associated temperatures and composition of smoke and gases. It calculates fire casualties as fatal injuries to occupants as they move about in the building in countering fire and smoke conditions. The EXITT module uses behavioral rules to predict the behavior of family groups. For example, if the fire does not block her route, a mother will attempt to rescue her child before leaving the building. The model is **deterministic,** meaning that given certain conditions or input, only a single result or output can occur. Subsequently, the people in the model either perform a predicted action or they do not. They are no calculations of the probability that a person will react in a certain way. As mentioned above in our discussion about how hard and soft systems approaches differ, such models yield deceptively specific and seemingly objective outputs. The next approach (CRISP) attempts to circumvent the specificity problem by expressing its outputs and probabilistic terms.

To recapitulate, HAZARD I is a **fire hazard assessment tool**; it is intended to represent the hazards associated with a single chosen fire scenario. Other types of models attempt to assess the risks associated with the probabilities that various types of fire could occur. The next section of this book discusses one such risk assessment that uses the predictions of human behaviors.

The Comparison of Risk Indices by Stimulation Procedures (CRISP) is a computer program fire risk assessment model that attempts to integrate building design, fire protection features in human behavior. CRISP was developed at the Fire Research Station in England. It includes supermodels that represent the physical and chemical processes of fire along with the behaviors of people attempting to escape or suppress the fire. CRISP models certain ways in which the physical environment affects people. For example, the fire generates carbon monoxide and smoke which hinders mental reasoning and vision. These attributes of the physical environment, in turn, inhibit people's ability to escape the fire. CRISP also models certain actions and rolled behaviors. For example, civilians can investigate and fight fires, worn and rescue other people, or simply escape on their own. CRISP is a **risk assessment model**, meaning that it yields probabilities that certain outcomes will transpire based on probabilities that certain types of fires will happen. The model generates probabilities by using a Monte Carlo approach. This means that the computer program randomly generates a limited sample of fire scenarios to estimate the risk to life of all possible fires. As an example, it chooses family members, and thus their associated behaviors, from a list of typical family combinations. Subsequently, the CRISP model differs from all the models that are deterministic, such as HAZARD I. Considering the great uncertainties about how fires in people behave in the real world, probabilistic models represent reality more accurately. However, as a rule, it is difficult or impossible to obtain probabilities that accurately represent the-real-deal fire. Of significance, the mathematics and computer programs for probabilistic models are more complex and costly to develop and use.

One of the features that I found interesting while researching for this book is that the CRISP model uses an approach in where the people in the program pursue various goals. Their behavior differs depending on whether they intend to warn others, fight the fire, or search for egress. Perhaps John Jay College professor Norman Groner said it best when he said, "It is difficult to predict and impossible to understand the behaviors of people without some knowledge of the goals that they are pursuing." This applies not only to people caught in fires but also to the people who design buildings and to those who write and enforce fire safety and building codes. The following sections of the book will discuss models based on goals.

Goal-Based Approaches

Fire safety design stems particularly from the engineering field, whereas soft systems models exist simply in the management sciences and in business schools. As noted earlier in this tax, the ability of soft systems approaches to handle a great variety of data types and interactions among systems components can prove very valuable. Of significance, this attribute makes them particularly useful for dealing with human behaviors. The Goal Based Models are exceptions to the generalization that fire safety analysis rarely used his soft systems approaches. **Goal-Based**

Models are attractive for a number of reasons. They can integrate both hardware and human systems elements usefully. For example, either automatic hardware or an emergency procedure can accomplish the goal of shutting a door and a fire rated barrier. Also, goals are useful as motivators of human performance. We can measure human performance in terms of meeting goals, for example, whether occupants evacuate the building within a certain time and whether a floor warden is also appointed and on site to assist in the evacuations. For the purpose of discussion this text breaks goal based models into two basic categories. The two goal based models are life safety strategies and goal decomposition approaches.

The building design and its occupants should work toward a common fire safety goal. A quick and total evacuation is the logical goal in many buildings but what about high rise buildings, hospitals, prisons and the myriad of other occupancy types were a quick and total evacuation is not possible or desirable? In this part of the text, you will learn about the concept of a life safety strategy as a goal towards which the performance of the buildings and occupants can work together as one system.

Life safety strategies are simple statements about the activities people should undertake in the event of a fire. The entire emergency plan may be quite complex, with different persons responsible for different actions but its overall organization reflects the goal of supporting one or more life safety strategies. At example of a life safety strategy for a high-rise might be when the alarm sounds, it alerts building occupants that they may need to evacuate. The building fire safety Director notifies floor wardens by emergency telephones when their respective floors should evacuate using which stairways. Obviously, the plan needed to make this life safety strategy work would be complicated, yet the strategy itself is just a simple statement around which to write the plan. There are two general reasons for writing life safety strategies. The first reason is so you don't get caught up in details that you neglect priorities. In general, is very easy to get caught up in details of building designs and features and emergency plans while losing sight of the overall approach. Ever hear the expression it's the old problem of not seeing the forest for the trees. Fire Service representatives to often make the same recommendations for almost any building without analyzing their appropriateness. The life safety strategy concept is a simple approach useful for tailoring fire emergency plans to specific buildings. The second reason for life safety strategies is that they help you communicate with people who are not knowledgeable about fire safety. Life safety strategies help facilitate communication among the people primarily concerned with building design like architects, fire protection engineer and code writers as well as the people who must use a building after its occupied. Building owners and managers are less resentful and more willing to comply with a standard when the rationale is explained by its contribution to a life safety strategy. When they understand that in an operative door closer is not simply a code violation but a serious threat to the life safety strategy of the building, they are more likely to do something about it. When the building manager and fire service have agreed on a life safety strategy, the fire service will be prepared to rescue people from threatened areas of refuge. Fire Service personnel will always give appropriate advice to occupants because they will know the best life safety strategy for every given building. Of significance, life safety strategies can be used to convince a building owner to exceed minimum standards and should be exceeded when building designs differ significantly from the typical structures and occupants around which the codes were conceived. Fire prevention inspectors, safety and engineering staff can set their priorities according to the importance of their activities to the life safety strategy for the building. I have seen safety staff conduct frequent checks of fire extinguishers and exit signs while fire doors remained propped open and sprinkler heads painted over.

Different types of buildings and occupancies often require different types of strategies. Life safety strategies often include certain basic features. However, you must know that these features are not mutually exclusive; that is, any single strategy can combine two or more of the following features. These features include but are not limited to; **occupants will simply use the nearest exit**, **refuge or staging areas** where building occupants are provided with a high degree of protection, **queued egress** or when building occupants wait before they can enter crowded stairs, **sequential egress** or when occupants in certain building areas (where they are in greater danger) evacuate before those in other areas, **localized egress** or when occupants in certain building areas (where they are in greater danger) evacuate while those and other well protected parts of the building simply remain in place, and **defend in place** or when building occupants are not moved at all an approach t know hat is common in healthcare facilities where patients can be moved only after much preparation.

Can you write the letter that matches the appropriate definition with each of the following features of life safety strategies?

Figure 3-1

_____ Localized egress	A. Building occupants are not moved at all. This approach is commonly employed in healthcare facilities where patients can be moved only after much preparation and with considerable assistance.
_____ Queued egress	B. Occupants in certain building areas, where they are in greater danger, evacuation, while occupants and other well protected parts of the building simply remain in place.
_____ Refuge or staging areas	C. Building occupants wait before they can enter crowded stairs.
_____ Sequential egress	D. Occupants in certain building areas, where they are in greater danger, evacuation before those in other areas.
_____ Defend-in-place	E. Certain building areas that provide building occupants with a high degree of protection.

The **goal-decomposition method** is simply a systematic way of looking at almost any type of decision making problem. Moreover, it is an approach that people commonly used in their every day lives and so it will seem sensible and familiar to most students. The goal-decomposition approach described here rests on methods of "cognitive task analysis" that are used to describe the "functional domains" within which people operate. Goal-decomposition yields a gold means network that describes the precise relationship among goals. In a gold-means network, the means for achieving higher level goals become goals in their own right and the means to pursue them then are described. Subsequently, a hierarchy network is created. The levels of the gold-means network are interrelated by a why—what—how logic. Ultimately, each higher level goal is a reason for pursuing goals at the next lower level, and each low a level goal is the means by which the goal at the next highest level is pursued. Simply put, goal "A" is the reason for pursuing goal "B", and goal "B" is the means by which goal "A" is pursued.[20]

Performance-based code approaches under development in the United States and Canada used the goal-decomposition approach. In the Canadian objective-based code approach, requirements are organized around a framework which clearly states the intent (objective) of each code requirement and then relates each of these objectives to higher and subsequently top level, objectives of the code document. In the United States, the proposed performance-based option for the NFPA 101 also relies on goal-decomposition, specifying life safety objectives and establishing performance criteria to verify an achievement of the objectives.

Decomposing goals is a useful approach to analyzing systems performance because it can include anything that affects performance. All types of performances of both people and hardware systems can fit easily in the same system analysis. This is not true for hard systems analysis because they can handle only specific types of quantifiable relationships. Of significance, is the importance in regards to human performance because the manner in which people behave in preparation for and during fires is difficult to predict. Subsequently, in goal-decomposition it is not necessary to predict behavior accurately. Instead, the purpose of a goal-decomposition is to decide what those responsible can do to control systems performance, including human behavior. After all, the ultimate purpose for fire safety design is to control the performance of fire safety systems. Moreover, it is useful to note that fire protection engineers successfully designed fire sprinklers in areas that stop the spread of fire and smoke before they had the analytical tools to measure the performance of their desires accurately.

The **Fire Safety Concepts Tree** is an example of a hierarchy decomposition of goals. Like the scenario-based goal-decomposition approach, it is an intentional systems representation of fire safety problem. A special NFPA committee on systems concepts developed this approach. NFPA 550, titled Guide to the *Fire Safety Concepts Tree*, discusses this approach, which is used wildly in the fire protection field. The fire safety concepts tree resembles an event tree. The most commonly used type of event tree is a fault tree. Fault trees are hard systems models that trace the potentially negative consequences of any particular failure selected for analysis. The fire safety concepts tree differs from a fault tree in important ways. In the fire safety concepts tree, the paths in the tree lead to success rather than failure. In an event tree of any type, the elements are always specific events with a designated probability of occurrence. However, the fire safety concepts tree is much more open ended. The elements in the fire safety concepts tree initially lacked definition. As the branches of the tree a merge, the concepts become increasingly specific and measurable. In effect, the fire safety concepts tree is an example of a goal-decomposition mythology.

As with other goal-decomposition methods, the decomposition of system states appears graphically as a gold-means network. As the network of goals undergoes decomposition, the associated system states become more concrete and measurable. At higher levels in a network, an example of a goal might be a systems state where all occupants not intimate with the fire are protected from dangerous levels of products of combustion. An example of system states for low-level goals might be all occupants are evacuated from the building within five minutes of the alone sounding and no potential fire exceeds 200°F. These goals, in turn, can be decomposed to increasing levels of specificity.

Figure 3-2

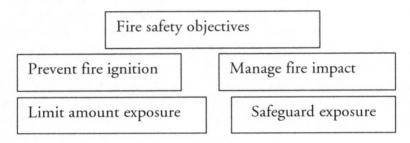

Uppermost Branches of the Fire Safety Concepts Tree

A Comprehensive Look at the World Trade Center Evacuation

The World Trade Center (WTC) complex was built for the Port Authority of New York and New Jersey. Construction began in 1966 and tenets began to occupy its space in 1970. The Twin Towers came to occupy a unique and symbolic place for both the Nation and Citizens of New York City. The World Trade Center actually consisted of seven buildings, including one hotel, spread across 16 acres of land. The buildings were connected by an underground mall known as the concourse. The Twin Towers or number one World Trade Center which was the North Tower and number two World Trade Center which was known as the South Tower were the signature structures, containing 10.4 million square feet of commercial office space. Both towers had 110 stories, or about 1,350 feet high and were square; each wall measured 208 feet in length. On any given workday, up to 50,000 office workers occupied the towers and 40,000 people passed through the complex.[21]

Each tower contained three central stairwells, which ran essentially from top to bottom and 99 elevators. Generally, elevators originating in the lobby ran to "sky lobbies" on higher floors, we're additional elevators carried passengers to the tops of the buildings. The stairwells A and C ran from the 110th floor to the raised mezzanine level of the lobby. Stairway B ran from the 107th floor to the level B6, six floors below the ground and was accessible from the West Street lobby level, which was one floor below the mezzanine. All three stairwells ran essentially straight up and down, except for two deviations in stairwells A and C where the staircase jutted out towards the perimeter of the building. On the upper and lower boundaries of these deviations was transfer stairways connected within the stairwell proper. Each hallway contains smoke doors to prevent smoke from rising from lower to upper portions of the building; they were kept closed but not locked. Doors leading from tenant space into the stairwells were never kept locked; reentry from the stairwells was generally possible on at least every fourth floor. Doors leading to the roof were locked. There was no rooftop evacuation plan. The roof's of both the North and South Tower was sloped and uncluttered surfaces with radiation hazards, making them impractical for helicopter landings and as staging areas for civilians. Although the South Tower roof had a helipad, it did not meet 1994 Federal Aviation Administration Guidelines.[22]

The first response on September 11, 2001 came from private firms and individuals, the people and companies in the World Trade Center complex. Everything that what happened to them going the next few minutes would turn on their circumstances and their preparedness, assisted by building personnel on-site. Of significance, we must continue to study their behavioral response to the subsequent fire if we are to gain anything from these tragic events.

Hundreds of civilians were trapped on or above the 92nd floor and gathered in large and small groups, primarily between the 103rd and 106th floors. A large group was reported on the 92nd floor, technically below the impact but unable to dissent. Civilians were also trapped in elevators. Other civilians below the impact zone, mostly on floors in the 70s and 80s but also on at least the 47th and 22nd floors, were either trapped or waiting for firefighters to save them.

Following protocol, the North Tower's fire safety director gave announcements to those floors that had generated computerized alarms, advising those tenants to dissent to points of safety. Preplanned fire drills called for tenants to descend to at least two floors below the smoke or fire and wait there for further instructions. Because of damage to the building's systems public address announcements were not heard in many locations. For this reason many civilians were not able to use the emergency intercom phone, as they had been advised to do and fire drills. The 911 operators and FDNY dispatchers had no information about either the location or the magnitude of the impact zone and were therefore unable to provide information as fundamental as whether a caller was above or below the fire. Because the operators were not informed of NYPD aviations determination of the impossibility of rooftop rescues from the point in towers on that day, they could not knowingly answer when callers asked whether to go up or down. Subsequently, 911 operators and FDNY dispatchers relied on standard operating procedures for high-rise fires. Therefore, civilians should stay low, remain where they are and wait for firefighters to reach them. This advice was given to callers from the North Tower for locations both above and below the impact zone. Although the guidance to stay in place may seem understandable in cases of conventional high-rise fires, FDNY Chief Officers in the North Tower lobby determined at once that all building occupants should attempt to evacuate immediately. By 8:57 a.m., FDNY Fire Chiefs had instructed the Port Authority Police Department and WTC Building Personnel to evacuate the North and South Tower as well because of the magnitude of the damage caused by the first planes impacts. Of significance, these critical decisions were not conveyed to 911 operators or to FDNY dispatchers. Moreover, in parting from protocol a number of operators told callers that they could break windows and several operators advised callers to evacuate if they could.[23]

Ultimately, most civilians who were not obstructed from proceeding began evacuating without waiting for instructions over the intercom system. Some remained to wait for help, as advised by 911 operators. Others simply continue to work or delayed to collect personal items but in many cases were urged to leave by others. While evacuating, some civilians had trouble reaching the exits because of damage caused by the impact. Some were confused by deviations in the increasingly crowded stairwells and impeded by doors that appear to be locked but actually were jammed by the breeze or shifting that resulted from the impact of the plane. Despite these obstacles, the evacuation was relatively calm and orderly. It is estimated that within 10 minutes of impact, smoke was beginning to rise to the upper floors in debilitating volumes and isolated fires were reported, although there were some pockets of refuge. Faced with insufferable heat, smoke, and fire, and with no prospect for relief, some jumped from the building.

Many civilians in the South Tower were initially unaware of what had happened in the other tower. Some believed an incident had occurred in their building; others were aware that a major explosion had occurred on the upper floors of the North Tower. Many people decided to leave and some were advised to do so by fire wardens. In addition, Morgan Stanley, which occupied more than 20 floors of the South Tower, evacuated its employees by the decision of company security officials. Consistent with protocol, at 8:49 a.m. the deputy fire safety director in the South Tower to hold his counterpart in the North Tower that he would wait to hear from "the bosses from the fire department or somebody" before ordering an evacuation. At about this time, an announcement over the public address system in the South Tower stated that the incident had occurred in the other building and advised tenants, generally, that their building was safe and that they should remain on or return to their offices or floors. A statement from the deputy fire safety director informing tenants that the incident had occurred in the other building was consistent with protocol; the expanded advice did not correspond to any existing written protocol and did not reflect any instruction known to have been given to the deputy fire safety director that day. We do not know the reason for the announcement, as both the deputy fire safety director believed to have made it and the director of fire safety for the World Trade Center complex perished in the South Palace collapsed. Clearly, however, the prospect of another plane hitting the second building was beyond the contemplation of anyone giving advice. Accordingly to one of the first fire chiefs to arrive, such a scenario was unimaginable, "beyond our consciousness." As a result of the announcement, many civilians remained on their floors. Others reversed their evacuation and went back up. Similar advice was given in person by security officials in both the ground floor lobby, we're a group of 20 that had descended by the elevators was personally instructed to go back upstairs, and in the sky lobby, where many waited for express elevators to take them down. Security officials who gave this advice were not part of the fire safety staff. Several South Palo occupants called the Port Authority police desk located in five World Trade Center and were advised to standby for further instructions; others were strongly advised to leave. It is not known whether the order by the FDNY to evacuate the South Tower was received by the Deputy Fire Safety Director making announcements there. However, at approximately 90 2 a.m. <minute before the building was hit in instruction over the South Tower's public address system advise civilians, generally, that they could begin and orderly evacuation if conditions warranted. Like the earlier advice to remain in place, it did not correspond to any pre-written emergency instructions.[24]

In the 17 minutes between 8:46 a.m. and 9:03 a.m. on September 11, 2001 New York City had mobilized the largest rescue operation in the city's history. Well over a thousand first responses had been deployed, and evacuation had begun in the critical decision that the fire could not before it had been made. Of significance and detriment, then the second plane hit.

At the lower end of impact in the South Tower, the 78th floor sky lobby, hundreds had been waiting to evacuate when the second plane hit. Many had attempted but failed to squeeze into packed express elevators. Upon impact, many were killed or severely injured; others were relatively unharmed. We know of at least one civilian who seized the initiative and shouted that anyone who could walk should walk to the stairs, and anyone who could help should help others in need of assistance. As a result, at least two small groups of civilians descend from that floor. Others remained on the floor to help the injured and move victims who were unable to walk to the stairwell to aid their rescue. Still others remained alive in the impact zone above the 78th floor. Damage was extensive, and conditions were highly precarious. The only survive unknown to have escaped from the heart of the impact zone described the 81st floor, where the wing of the plane had sliced through his office, as a demolition site in which everything was broken up and the smelled of jet fuel was so strong that it was almost impossible to breeze. This person and escaped by means of an unlikely rescue, aided by a civilian fire warden descending from a high of floor, who, critically, had been provided with a flashlight. At least four people were able to descend stairway A from the 81st floor or above. One left the 84th floor on immediately after the building was hit. Even at that point, the stairway was dark, smoky, and difficult to navigate; glow strips on the stairs and handrails were a significant help. Several flights down, however, the vacuum he became confused when he reached a small door that caused him to believe the stairway had ended. He was unable to exit that stairwell and switch to another. Many civilians in and above the impact zone ascended the stairs. One small group reversed its dissent down stairwell A after being advised by another civilian that they were approaching a floor in flames. The only known survivor has told us that their intention was to exit the stairwell in search of clearer air. At the 91st floor, joined by others from intervening floors, they perceive themselves to be trapped in the stairwell and began descending again. By this time, the stairwell was pretty black, intensifying smoke caused many to pass out and fire had ignited in the 82nd floor transverse hallway. Others ascended to attempt to reach the roof but were thwarted by locked doors. At approximately 9:30 a.m. a lock release order which would unlock all areas in the complex controlled by the buildings computerize security system, including doors leading to the roof, was transmitted to the security command Center located on the 22nd floor of the North Tower. Damage to the software controlling the system, resulting from the impact of the plane, prevented this order from being executed. Others, attempting to descend, were frustrated by Jan or locked door is in stairwells or confused by the structure of the stairwell deviations. By the lower 70s, however, the stairwells A and B were well-lit and conditions would generally normal. Some civilians remained unaffected floors and a least one ascended from a low point into the impact zone, to help evacuate colleagues or assist the injured. Within 15 minutes after the impact, debilitating smoke had reached at least one location on the hundredth floor, and severe smoke conditions were reported throughout the floors in the 90s and hundreds over the course of the following half hour. By 9:30 a.m. in number of civilians who had failed to reach the roof remained on the hundred and fifth floor, likely unable to descend because of intensifying smoke in the stairwell. There were reports of tremendous smoke on that floor but a least one area remained less affected until shortly before the building collapsed. There were several areas between the impact zone in the uppermost floors where conditions were better. At least a hundred people remained alive on the 88th and 89th floors, in some cases calling 911 for direction.[25]

Evidence suggests that the public address system did not continue to function after the building was hit. A group of people trapped on the 97th floor, however, made repeated references in calls to 911 to having heard "announcements" to go down the stairs. Evacuation tones were heard in locations both above and below the impact zone. By 9:35 a.m., the West Street lobby level of the South Tower was becoming overwhelmingly crowded by injured people who had descended to the lobby but could not physically go on from there. Those who could continue with direct it to exit north or east through the concourse and then out of the World Trade Center complex. By 9:59 a.m., at least one person had descended from as high as the 91st floor of the South Tower and stairwell A was reported to have been almost empty. Still will be was also reported to have contained only a handful of descending civilians at an earlier point in the morning. But just before the tower collapsed, a stream of civilians had been observed descending a stairwell in the 20s. It is believed that these civilians were descending from at or above the impact zone.[26]

In the North Tower, civilians continued their evacuation. On the 91st floor, the highest floor with stairway access, all civilians but one, were uninjured and able to descend. While some complained of smoke, he, fumes, and crowding in the stairwells, conditions were otherwise fairly normal on floors below the impact. At least one stairwell was reported to have been clear and bright from the upper 80s down. Those who called 911 from floors below the impact were generally advised to remain in place. One group trapped on the 83rd floor pleaded repeatedly to know whether the fire was above or below them, specifically asking if 911 operators had any information from the outside or from the news. The callers were transferred back and forth several times in advised to stay put. Evidence suggests that these callers died. At 8:59 a.m., the Port Authority Police Desk at Newark Airport told a third party that a group of Port Authority civilian employees on the 64th floor should evacuate. The third party was not at the

World Trade Center but had been in phone contact with the group on the 64th floor. At 9:10 a.m., in response to it in query from the employees themselves, the Port Authority police desk in Jersey City confirmed that employees on the 64th floor should be careful, stay near the stairwells and wait for police to come up and get them. When a third party inquired again at 9:30 a.m., the police desk at Newark Airport advised that they absolutely evacuate. The third party informed the police desk that the employees had previously received contrary advice from the FDNY, which could only have come from 911. These workers were not trapped, yet unlike most occupants on the upper floors, they had chosen not to descend immediately after impact. They eventually began to this end the stairs but most of them died in the collapse of the North Tower. All civilians who reached the lobby was directed by both NYPD and PAPD officers into the concourse, where although police officers guided them to exit the concourse and complex to the north and east so that they might avoid falling debris's and victims. By 9:55 a.m. only a few civilians were descending above the 25th floor and stairwell B; these primarily were injured, handicapped, elderly, or severely overweight civilians, in some cases being assisted by other civilians. By 9:59 a.m. tenants from the 91st floor had already descended the stairs and exited the concourse. However, a number of civilians remained in at least stairwells C, approaching lower floors. Other evacuees were killed earlier by the debris falling on the street.[27]

The 911 calls placed from most locations in the North Tower grew increasingly desperate as time went on. As late as 10:28 a.m., people remained alive in some locations, including on the 92nd and 79th floors. Below the impact zone, it is likely that most civilians who were physically and emotionally capable of descending at exited the tower. The civilians who were nearing the bottom of stairwells see were assisted out of the building by FDNY, NYPD and PAPD personnel. Others who experienced difficulty evacuating were being helped by first responders on lower floors.

The emergency response to the attacks on 9/11 was ultimately improvised. In New York City, the FDNY, NYPD, PAPD, WTC employees, and the building occupants themselves did their best to cope with the effects of an unimaginable catastrophe which unfolded furiously over a mere 102 minutes for which they were unprepared in terms of both training and mindset. As a result of the efforts of first responders, assistance from each other, and their own good instincts and goodwill, the vast majority of civilians below the impact zone were able to evacuate the towers. The National Institute of Standards and Technology (NIST) provided a pulmonary estimation that between 16,400 and 18,800 civilians were in the World Trade Center complex as of 8:46 a.m. on September all 11. At most 2,152 individuals died at the World Trade Center complex who were not fire or police first responders, security or fire safety personnel of the World Trade Center Complex or individual companies, volunteer civilians who ran to the World Trade Center after the planes impact to help others, or on the two planes that crashed into the twin towers. Out of this total number of fatalities, we can account for the workplace location of 2,052 individuals or 95.35%. Of this number 1,942 or 94.64% either worked or was supposed to attend the meeting at or above the respected impact zones of the twin Towers; only 110 or 5.36% of those who died, worked below the impact zone. While a given person's office location at the World Trade Center does not definitely indicate where the individual died that morning or whether he or she could have evacuated, this data strongly suggest that the evacuation was a success for civilians below the impact zone. Several factors influenced the evacuation on September 11. It was aided greatly by changes made by the Port Authority in response to the 1993 bombing and by the training of both Port Authority personnel and civilians after that time. Stairwells remained lit near unaffected floors; some tenants relied on procedures learned in fire drills to help them to safety; others were guided down the stairs by fire safety officials based in the lobby. Because of damage caused by the impact of the planes, the capability of the sophisticated building systems may have been impaired. Rudimentary improvements, however, such as the addition of glow strips to the hand rails and stairs, were credited by some as the reason for their survival. The general evacuation time for the towers dropped from more than four hours in 1993 to under one hour on September all 11 for most civilians who were not trapped or physically incapable of enduring a long dissent. First responders played a significant role in the success of the evacuation. Some specific rescues a quantifiable, such as an FDNY company rescue of civilians trapped on the 22nd floor of the north tower, or the success of FDNY, PAPD, and NYPD personnel in carrying non-ambulatory civilians out of both the North and South Towers. In other instances, intangibles combine to reduce what could have been a much higher death toll. It is impossible to measure how many more civilians who descended to the ground floors would have died but not for the FDNY, NYPD and PAPD personnel directing them via safety exit routes to avoid jumpers and falling debris, to leave the complex urgently but calmly. It is impossible to measure how many more civilians would have died but for the determination of many members of the FDNY, PAPD, and NYPD to continue assisting civilians after the South Tower collapsed. It is impossible to measure the calming influence that a sending firefighters had on the sending civilians or whether but for the firefighters presents the poor behavior of very few civilians could have caused a dangerous and panicked mob like mentality of fight and flight. Of significance the positive impact of first responders on the evacuation came at a tremendous cost of firefighters and police officers lives.[28]

The first responders on 9/11, as in most catastrophes, were private-sector civilians. Be cool is 85% of our nation's critical infrastructure is controlled not by government but by the private sector, private-sector civilians are likely to be the first responders in any future catastrophes. For that reason, we must continue to assess the state of private sector and civilian preparedness in order to formulate recommendations to address this critical need. Ultimately, it can be said that all future fire-related behavioral research will grow out of experience of the civilians at the World Trade Center on 9/11.

An Evacuation Model for High-Rise Buildings

EXIT89 is an evacuation model designed to handle the evacuation of a larger population of individuals from a high-rise building. It can track the location of individuals as they move through the building so that the output from this model can be used as input to a toxicity model that will accumulate occupant exposures to combustible products. The model allows the user to specify whether the occupants of the building will follow the shortest exit paths or their familiar route from the building, as well as to allow evacuation delays to be set by the user by locations and additional delays to be distributed randomly among the occupants. It allows smoke input to be read in from smoke movement model or from user defined blockages. Additionally, this program accounts for occupant density in building spaces to compute each occupants walking speed. The model described in this textbook was designed to use the smoke movement data generated by one component of the HAZARD I program and to provide the occupant location data required by the ability model incorporated in HAZARD I. Moreover, this program has been tested using data from evacuation drills in several buildings.

EXIT89 was designed to model the evacuation of a large building with the capability of tracking each occupant individually. The output of this model, in combination with a fire and smoke movement model using the same building layout, can be used to predict the effects of cumulative exposure to the toxic environment often found in high-rise fires. Ultimately, it differs from other large population evacuation models, such as network flow models, that treat the occupants of a building as if they were a fluid in a pipeline. Although, models are capable of predicting points of congestion and time until areas are cleared of occupants, they cannot follow the movement of individuals separately. Additionally, EXIT89 differs from evacuation models that incorporate specific occupant behaviors because the size of the population that can be handled by EXIT89 is too large to handle such a large amount of detail. Behaviors can be implicitly modeled to some degree by using some of the features recently added to the model. Delays in beginning evacuation are common in real situations, where occupants may assume that they are hearing another false alarm, or they may hesitate to respond to clues, including smoke, because no one else is reacting. Delays can also occur as a result of activities the occupants engage in before beginning to exit the building. These delays can include investigating the source of the alarm or smoke, securing files, gathering personal belongings, and notifying others of the situation. These delays can now be incorporated by setting a delay to each location in a building and having all occupants at the location wait that amount of time before beginning to leave. Data from real evacuation's have also shown that delays occurred going the course of exiting the building as people seek information, gather belongings, alert others, and fight the fire. As a first step toward simulating that occurrence, the ladies can now be randomly assigned to any specified proportion of the occupants of the building. Observations of exit choices during evacuations does indicate that occupants of a building will often take the same route out of the building that they took coming in. Of significance, EXIT89 was modified to allow the user to model his behavior, rather than have all occupants follow calculated shortest routes out of the building.

The remainder of this chapter gives a brief description of the model and its components as well as three example applications are presented to illustrate the use of these and other user options recently added to the EXIT89 model. Ultimately, this model is another example of the triumphs that have grown out of civilian experiences at the World Trade Center Complex on September 11. Much useful information derived out of this model will save lives in the future and is a credit to the National Institute of Standards and Technologies Building and Fire Research Laboratory Grant Program.

The program requires as input a network description of the building, geometrical data for each room and for openings between rooms, the number of occupants located at each node throughout the building, and smoke data if the effect of smoke blockages is to be considered. The user is allowed to select among several options, including whether the occupants of the building will follow shortest paths out of the building or will use familiar routes; whether smoke data, if any, comes from a fire and smoke model or will be input as blockages by the user; whether there are any delays in evacuation throughout the building; whether there are any additional delays in evacuation among the occupants of the building and, if so, what percentage of the occupants will delay and what are the minimum and maximum times for delay.

The model either calculates the shortest route from each building location to a location of safety, usually outside, or sets user defined routes throughout the building. It moves people along the calculated or defined routes until a location is blocked by smoke. Affected exit routes are recalculated and people movement continues until the next blockage occurs or until everyone who can escape has reached the outside. Evacuation can begin for all occupants at time zero or can be delayed. Additionally, delays over a specified range of time can be randomly assigned to occupants. Smoke data can be used to predict when the evacuation of a smoke detector would occur and evacuation will begin then or after some user defined delay beyond that time. Of significance, the program is written in FOR-TRAN and currently runs on an IBM mainframe.

The model was designed to meet the following needs for use in high-rise applications. The model was made to handle a large occupant population; to be able to recalculate exit paths after rooms or nodes become blocked by smoke; to track individuals as they move through the building by recording each occupants location at set time intervals during the fire; and to fairy travel speeds as a function of the changing crowdedness of space is touring the evacuation, an example of **queuing effects** (when a number of people wait in line, or in queue, before entering the exit) which were previously described in this text.

The size of the building and its population that can be handled by this program can be expanded by modifying the size of the data arrays used by the program the dimensions of the storage arrays currently allow for over 700 occupants in a total of over 300 nodes are building space is during 100, ten second time intervals. These can be changed by the user to handle larger problems. Due to the naming convention of nodes that the program relies on, each floor can have up to 89 nodes in the building can have up to 10 stairways.

It must be mentioned that the model does not have global perspective, that is, occupants are not directed along the truly shortest path out of the building. When the shortest path option is selected, people will move to the closest exit on a floor, even though the total length of the path to the outside might be shorter if another exit were used. For example, a guest of a high-rise hotel might step out of his room and head to the closest stairwell even though it may be five flights down to grade level while another stairwell a slightly greater distance from his room might be only three flights from grade level. A model with a global perspective would move him along the truly shortest path, but that route would not be realistic for a hotel occupant who would be unfamiliar with the layout of a building. The option that allows the user to specify that occupants will follow familiar routes can be used to model the situation where, for example, staff will know that a certain route is shorter, or a more likely case, where people will travel out the route they followed on entering the building. If they smoke blockages occur is going evacuation, the recalculation of routes for a floor will use the shortest route algorithm discussed below. Another assumption is that once people enter a stair-well, they will follow it all the way down to the outside unless it becomes blocked by the fire's progress, in which case they will move out of the stairs and onto the nearest floor. In real situations, people may head for the roof or leave the stairs to go onto lower floors for no apparent reason.

Unfortunately, the program does not explicitly include any fire related behavioral considerations. These behavioral considerations include but are not limited to the investigation of the fire, rescue of people, and alerting or assisting occupants who may require help. Mainly, the population of a high-rise building is too large to handle so much detail for each individual. Moreover, many behavioral researchers do believe that behaviors such as investigation or rescue of other occupants are not as relevant in larger, more impersonal buildings. Subsequently these behaviors can cause delays in evacuation in these behaviors can be included implicitly by using the option to add delays to all occupants at various locations and then adding additional delays to randomly select individuals.

How does the model calculate walking speed? The model calculates walking speed as a function of density. This function of density is based on formulas from Predtechenskii and Milinskii. Body size is included in their density calculations. Using dimensions of people in various types of dress, both empty-handed or income bird with packages, knapsacks, big age or babies, they calculated the area of horizontal projection of a person. This measurement is the area of an ellipse whose axis corresponds to the width of a person at shoulder level and breath at chest level.[29]

As-far-as the model goes a node is defined as a room or sections of a room or corridors, which ever will result in the most realistic travel paths.

The definition of each node includes its usable floor area, the height of the ceiling, its initial occupant load, the number of seconds occupants of that room will delay before being evacuated, and the node an occupant will move to if the user chooses the option of having occupants move along defined routes. Within the program is defined arc's, which are the distances between the nodes and the width of the opening between the nodes. Arcs are bidirectional so a connection between nodes only has to be described once. Escaping by a window is allowed by assigning a very large value as the distance along the arc so that the route will only be used only as a last resort.

There are six options set by the user at the beginning of the input file. The first indicates whether metric or standard measures will be used in input and output. Of significance, it should be noted that all calculations are done in metric scale but this option allows the simple use of evacuation data and floor plans from a variety of sources. The second option specifies the body size used as the basis of density calculations that are used to calculate velocities. These choices are extremely sophisticated and will not be covered in this textbook. The third option allows the user to specify whether occupants will be moving at emergency or normal (slower) speeds. This will also not be described in this textbook. The fourth option allows the user to determine whether the program should calculate the shortest paths between the nodes or whether the user will be specifying the node to which occupants will move from each node. If the user selects specified routes, the node to which occupants of a node will move is included as part of the node description in the input. User-specified pass will be used until a node on a floor becomes blocked by smoke. In that case, the routes for the floor will be recalculated using the shortest route. The fifth option indicates whether or not the user is reading in smoke data from a program known as CFAST or whether there will be user-defined blockages or no blockages. The use of this option will also not be described in this text. Finally, the user selects the full output, which prints information every time someone moves from one space to another or summary output showing floor and stairway clearing times and usage of exits.

Lastly the user indicates whether or not additional delay times should be randomly distributed among the occupants. If yes, the user then specifies for what percentage of the occupants there will be additional delays and over what range of time usually in seconds those delays should be chosen.

To recapitulate, the model in its current form does not include any explicit behavioral considerations but it does allow behavioral considerations to be handled implicitly by incorporating time to perform investigation activities or to alert others before evacuating in the delay times that the user specifies for the occupants of each node. In addition to specifying delay times for each location, the user now can also have the computer randomly assign additional delays to some percentage of the individuals throughout the building. In this same way, another behavior that can be dealt with implicitly is the tendency of able-bodied adults in the presence of other able-bodied adults to ignore early warning signs of the presence of a fire. EXIT89 now allows the user to model the frequently observed tendency of occupants to follow the route out of a building that they are most familiar with, not the shortest paths out of the building which orphaned would involve the use of emergency exits. These familiar past defined by the user will remain in place until a location on that floor becomes blocked by smoke and the routes on that floor need to be recalculated using the shortest route algorithm. Walking speeds are calculated as a function of density and are based on tables of values from Predtechenskii and Milinskii. The model does not yet simulate crawling through the smoky rooms by reducing walking speeds or reversing direction where possible to use a less smoky, so long the escape route. Also to be included is the simulation of disabled people, who can be incorporated using added size, in order to impact density, as well as slower speeds. Of significance, the University of Ulster has provided much useful information in incorporating this modification. One of the programs inputs is the capacity nodes. The reason for including this value was to allow evacuees to avoid nodes that were already crowded if alternate routes are available. This would prevent occupants from queuing at one stairway while the other section or sections of the floor emptied out into less busy or stairways. Refinements of the program to define and possibly limit the range of a smoke detector also need to be added to the model. Future plans for the model include adding a component for disabled occupants and documenting from available literature travel speeds in the delay times that can be used for occupants to begin evacuation and for delays during evacuation. Subsequently, these travel speeds and delay times may be occupancy specific. Ultimately, testing of the model using data from actual emergency and non-emergency evacuations must continue.

Acknowledgments

This work has been sponsored in part by the National Institute of Standards and Technology Building and Fire Research Laboratory under grant number 60NANB2D1286.

Rita F. Fahy, *"EXIT89An Evacuation Model for High-rise Buildings,* Model *Description and Example Applications,"* Fire Analysis and Research Division National Fire Protection Association, 1 Batterymarch Park Quincy, Massachusetts 02269-9101 USA.

Private Dwelling Fires

Of significance, most fire fatalities in the United States of America occur in one and two family private dwellings. Why do you think this is so? Is there anything particularly dangerous about the structures as compared to other types of residential units, for example apartments? Moreover, people are much more likely to die in their own homes than in other structures. Why do you think this is the case? After trying to answer these questions, you should understand that neither the residential fire problem though the solution is simple and straightforward.

Staten Island New York is a typical borough of New York City. So typical, in fact, about 70% of its housing units are single-family detached or attached dwelling units, very close to the percentage of the country as a whole. For a number of years, the only fires that have coursed fatalities in Staten Island have occurred in these units. Staten Island has been lucky in the last few years; it had suffered no major fire fatalities. But the boroughs luck is running out. In the early morning hours of December 2005, an unusually tragic fire struck one of Staten Island's established low income neighborhoods. Any young girl and boy died, and a parent was severely burned trying to rescue them.

You were informed in Chapter 1 of this text, that economics are correlated with fire losses and casualties. The casual relationships are complicated; there are many ways in which low income status can lead to increased fire risk. However, in this narrative the problem is straightforward. Poorer families are more likely to leave children at home without adult supervision, not because they are less caring parents, but because finding adequate and affordable child care can be exceedingly difficult. Unsupervised children are at particular risk, both because they can start fires and because they are less likely to be able to escape.

When we know the events that lead to a fire incident, it can appear that the victims took excessive and unreasonable risk. There is research that demonstrates **hindsight bias**, meaning that people consistently exaggerate the predictability of outcomes. Looking back, events seem more inevitable than they really were. We tend to think that people could have foreseen easily that their actions will cause a fire. However, from the standpoint of the people involved, the associated decision can seem reasonable, and even necessary, at the time they made it.

You probably have had the opportunity of explaining something to younger children. As a firefighter, you may have visited an elementary school or preschool as part of your department's public education program. From the children's eager attention and questions, it probably appeared that they absorb the information that you provided. Yet if you ask questions, the amount of information the children really did not understand may have surprised you.

This December morning Sandy could not sleep, thinking of the upcoming Christmas holiday and began looking for something to do and notice candles. Watching a candle flame had always fascinated her. She found some matches that her mother had left on a table. She knew that her mother had forbidden her to light matches, but she had watched her mother so many times that she was sure that she could do it herself. She carefully placed a candle on her mother's bed, and struck the match. The match flared up, and she quickly lit the candle. She was so pleased with her success in fascinated by the candle flame that she didn't notice that the match was still lit when the match flame burned her finger, she involuntarily jerked her hand and knocked over the candle. The flame immediately ignited the loosely arranged betting and the laundry spew around the floor.

Young children have limited cognitive abilities that make it unlikely that they will understand concepts that, to an adult, may seem very simple. For example, pictures or video intended to illustrate the dangers of fire can actually increase children's interest in fire and inadvertently encourage them literally to play with fire. Public education efforts need to be appropriate to the age of the child. Young children will not understand safety messages beyond their cognitive abilities as they are intended. In this instance, Sandy understood the rule that forbid her to play with matches, but she did not understand the nature of the related danger. Without some sense of related danger, "not playing with matches" can be just another among the myriad rules told to a young child. Very possibly, Sandy may have felt that she was not playing with matches, what that she was lighting a candle. She certainly did not understand that the candle needed to be placed in Asia to a position well away from any combustible materials.

Sandy was terrified and thought only about hiding from the fire. She ran out of the bedroom and into her sleeping older brother Pedro's bedroom and closed the door. Within five minutes, the room fledged over and dense smoke filled the entire house.

You probably have heard stories, often true, about children trying to escape a fire by hiding under beds or in closets or bathrooms, even when they would have been able to escape had they taken a different action. This is another indication of their formative level of cognitive development. This particular type of mental model is an action schema. **Action schemas** are brief mental plans or scripts that certain situations evoke automatically. For example, when you go to the counter at the video store, you have an action schema for paying for your video rental. You take out your wallet, search for and take out an appropriate amount of money, hand the money to the checkout person and wait for change. You don't need to think about any of this, your response is to a familiar situation and automatically activates the sequence of actions. Unfortunately, regarding fire prevention adults have a greater repertoire of action schemas than children do. In this example, Sandy used in action schema that might be called hiding from a threat.

As an adult, you understand that this action schema is inappropriate when applied to fire. But as an adult, you find it much easier it to some an action schemas that are appropriate to a great range of situations. Education and training programs often emphasize the idea of having children practice an evacuation plan. Can you use the concept of action schemas to explain why this type of training is important? In what ways do you think children differ from adults in learning how to evacuate their homes?

Adults can understand and record a simple action schema at a verbal level; that is, they can listen to, understand, and remember a plan for escaping from their home. This is not necessarily true for young children. Many can learn this type of action schema only by actually performing the behaviors. Moreover, actually practicing an escape is an excellent way to learn in action schema.

Again, as we were informed in Chapter 1 this textbook they are gender differences in response to fires. Research has shown that women are more likely to alert others and help people out of the buildings than men; men are more likely than women to investigate and attack the fire.

Soon after this Staten Island fire started, a couple from the neighborhood saw flames through their bedroom window. They remembered that Sandy and Pedro were sometimes alone in the house late in the early mornings, as their mother finished late waiting tables at the diner. The husband a parent himself immediately tried the front door, which was locked. However, he quickly found an open window and climbed into the house. Meanwhile, his wife ran back to their house to call the fire department. Sue another neighbor joined in the efforts. He picked up rocks from the yard and threw them through the bedroom windows in an attempt to wake the children, with detrimental results. Tragically, both Sandy and Pedro died from smoke inhalation and massive burns. The neighbor who tried to rescue the unattended children suffered severe burns himself requiring intensive care from the Staten Island burn center but was eventually released. First arriving firefighters found the children still in the bedroom and the neighbor near the living room window unconscious.

The Uniqueness and Compromise of Residential Care Facility

The number of adult residential care facilities has increased rapidly during the last few decades, as a result of the increasing numbers of older people and the deinstitutionalization of persons with developmental disabilities and mental illness. For the purpose of this textbook **deinstitutionalization** refers to the historical trend in which, as a matter of public policy, persons with cognitive disabilities were moved out of large institutional settings, like mental hospitals and state schools. Ideally, these people would be moved into small, relatively normal settings, such as small group homes that resemble ordinary houses. Unfortunately, partially because of inadequate funding, many of these displaced people found themselves in unregulated and relatively unsafe settings, or even homeless. Only recently have codes recognized this new and different type of occupancy, and only some model codes have done so. Of significance, the National Fire Protection Association's Life Safety Code identifies these types of occupancies as "board-and-care homes," and the New York City Building Code identifies them as "adult residential care facilities." The NFPA Life Safety Code and the New York City Building Code identifies board-and-care or residential care facilities as a separate occupancy, and as a building or part thereof that is used for lodging and boarding of residents for the purpose of providing personal care services. Personal care specifically does not include medical or nursing care. Many types of facilities qualify to be board-and-care or residential care facilities including, but not limited to, halfway houses, assisted living facilities, residential care homes, and group homes. Residents of board-and-care or residential care facilities need help with certain activities because of their disabilities. However, these disabilities may or may not affect fire safety. For example, residents of a group home for mildly retarded persons may actually be more likely to survive a fire than the general public, because they are more likely to follow recommended procedures. However, residents of a group home for severely retarded persons may be entirely unable to protect themselves without assistance from staff members.

One important reason decisions about how to regulate these types of homes is so difficult is that societies are very complex systems. All regulations have unanticipated consequences. In the case of board-and-care homes or adult residential care facilities, the simplest solution was to apply healthcare codes to these facilities, thereby assuring a very high level off fire safety. However, these requirements were too expensive for many facilities to meet. Most of these homes are not very profitable because their residents have and receive very little money. Many operators of these facilities feel they cannot afford the safety upgrades required in the model codes. Subsequently, many are faced with unpleasant decisions. They could close their homes, sending their residents to live in places that did not meet any reasonable standard of fire safety and personal care, or they could operate illegally and hope that the authorities would look the other way. In general, regulations can affect social systems in complex and unexpected ways. Ironically, requiring the highest levels of fire safety actually can increase the overall fire risk by moving people into unregulated settings. Subsequently, in an attempt to find a happy medium the New York City Department of Buildings in conjunction with the Fire Department has adopted parts of the NFPA Life Safety Code and parts of the International Building Code to create a standard for adult residential care facilities. According to the New York City building code an adult residential care facility is a family type home which provides housing, supplies and services for adults who, though not requiring continual medical or nursing care, are by reason of physical or other limitations associated with age, physically or mentally disabled or other factors unable or substantially unable to live independently. Such facilities may be known as community residents, supervised community residents, supportive community residents or individual residential alternatives according to the degree of support they provide. In general these facilities are found in either one family or two family dwellings serving one to 14 persons or can be located in a multiple dwelling serving 15 or more adults.

As you learned in Chapter 1 of this textbook, socioeconomic factors are strongly related to the risk of becoming a fire victim. However, the relationships between socioeconomic status and fire risk are many and complex. For example, economic factors can play an important role in determining how stringent to set fire safety regulations and whether to comply with those regulations. Moreover, there is a point where fire safety become so expensive that some adult residential care facilities must either cease to operate or operate underground as unlicensed facilities. This is especially true where these types of facilities serve people on the margins of economic life, such as elderly persons without pensions or other persons with disabilities who receive little or no public assistance. State and local governments often have to make decisions involving trade-offs. Research on rational decision-making involves how many people can make the best possible decision. The term **normative decision-making** also is used in the research literature and has the same meaning as **rational decision-making**. Subsequently, much of the research on rational decision-making involves comparisons between alternatives based on how likely each decision is to lead to various desirable and undesirable outcomes. Of significance, many of the theoretical models used in rational decision-making closely resemble the methods used to conduct risk assessments in fire safety research. Moreover, this line of research has been very well developed over a period of several decades. You probably have made certain types of decisions in this fashion. For example, you have considered buying different sneakers based on how likely they were to achieve certain outcomes like how they look, if they are comfortable and how much they cost. However, researchers have found that people deviate from the ideal models of decision-making, even for the type of problem where clear alternatives lead to known out-comes. In my every day life I rarely encounter such well-defined problems but in the real world complexity and uncertainty characterized most of my decisions.

Fire related behavioral researchers have begun to focus on how people really make most of their decisions. Research on descriptive decision-making involves how people really make decisions in their everyday lives. In fact, some of the most important research on naturalistic decision-making has investigated how fire chiefs make key tactical decisions during firefighting operations. You may have occasionally had to make decisions that have become more difficult because of your role conflict. An example of role conflict is when different people expect different behaviors from the same role. A chief fire officer often feels conflict because his or her subordinates feel he or she should be their advocate and put their interests first. However, the fire chief is also a department head in city government, which carries the expectation that he or she will carry out city policy and graciously accept budget constraints even when not in the best interests of the fire department.

Mishandling of smoking material is the most frequent cause of fatal fires in these types of homes. Therefore, it was not unexpected when Mayor Mike Bloomberg the Mayor of New York City banned smoking altogether in these types of occupancies. However, operators of these homes have great difficulty controlling the smoking habits of residents. Residents of adult residential care facilities are not supervised as closely as residents of healthcare facilities. Also, smoking is an important quality of life issue for many residents, and may feel that they are entitled to smoke in their own home, regardless of health and fire risks. When residential care facilities bane smoking altogether the danger of fire actually can increase, since residents may try to sneak cigarettes in sleeping rooms or other areas. Of significance, it is especially easy for fires to start when resident's carelessly discards a cigarette or match because he or she is worried about being court.

In an adult residential care facility setting, it is inappropriate to use the healthcare model for emergency planning. From a system standpoint, the building is not constructed to provide long-term refuge in bedrooms and the number of staff is less than that found in healthcare settings. Accordingly, residents need to be trained to escape quickly, to the level that they are cognitive and physical abilities allow. To the degree that it is not possible to train them, planning needs to incorporate measures to compensate; examples include special staff procedures, flashing lights for persons with hearing impairments etc. It is common to fine at these types of homes, the owner, usually trained in healthcare settings, believes' that she or he is following the best possible procedure by following codes modeled after healthcare settings. Many fire companies upon fire prevention inspection often encounter and think that these types of occupancies are of an institutional type and are fooled to believing that the procedures if conformed to healthcare facilities are appropriate. Behavioral researchers labeled this form of thinking as **negative transfer,** where something learned in one situation actually interferes with learning new procedures.

The board and care occupancy or adult residential care facility is notable for a wide range of disabilities found among residents. It is important to note that a wide range of abilities and disabilities also exists among the general population. The great over representation of the very young and very old among fire casualties demonstrates the importance of human performance. For these types of facilities assessing human performance is important to planning for a fire or emergency. As part of its efforts to design a fire safety of evaluation system for these types of occupancies, the research team at NIST designed a means for conservatively estimating the level of difficulty involved in evacuation this type of home. When NFPA adopted its Life Safety Code it included this "evacuation difficulty index" as an optional method for classifying board-and-care homes into one of three levels of evacuation difficulty; they are prompt, slow, and impractical. Students can find this method in NFPA 101A, *Guide on Alternative Approaches to Life Safety.* Ultimately, making assumptions based on the mere existence of disabilities often can be too mistaken judgments about how well people can protect themselves during fires or emergencies. For example, a person with the sight impairment might have less difficulty finding his or her way out of a building filled with smoke because he or she would know how to navigate without having to see their way. But the need for a hearing aid can prevent people from detecting a fire alarm signal at night, putting them in grave jeopardy if assistance is not provided. The way that residents are classified can influence the way that fire safety regulations are applied. Of significance, state regulations simply classify residents in these types of facilities mostly as "mobile" or "immobile." However, the life safety code uses a more sophisticated approach that considers residents needs for assistance along with the availability of staff to provide for those needs. The combination of staff and residents really determines how quickly occupants can evacuate a facility. Therefore, the life safety code classifies facilities, not individual residents, as having evacuation capability ratings of "prompt," "slow," or "impractical." The "evacuation capability determination method" in NFPA 101A is designed specifically to evaluate people according to their expected performance during a fire or emergency. It does not consider the presence of a specific type of disability, in itself. It takes as relevant only the disabilities affect on fire safety performance. If you need to assess the impact of disabilities on fire safety in these types of homes or facilities, this assessment tool provides a good overview of the issue that you'll want to consider and is a direct result of dedicated fire-related behavioral research.

Vertical openings, particularly stairs have been implicated in several of these types of home fires. Many of these facilities are converted from large single-family houses that have attractive open stairs leading to the second floor. These can be enclosed only at considerable expense. Nonetheless, the fire service must recognize this hazard and require the enclosure. As with the importance of keeping the door closed to a fire area behavioral scientists and researchers can state this problem as not having an adequate mental model of fire protection basics. You can think of mental models as greatly simplified representations of more complex systems and the events in the real world. Many people do not have good mental models for fire protection. An important example is that they greatly on the estimate how quickly fire can grow and spread. Subsequently, there is another type of mental model that is important and fire safety that is "people's mental maps of building layouts." One of the reasons why people don't make better use of emergency exits is that the exit paths are not part of their mental maps. Ultimately, this is the most important reason for conducting fire exit drills.

Persons with disabilities often feel they should be allowed to assume a greater risk than the general public in exchange for equal opportunities and access. With us, many such persons feel that it is not legitimate to deny them access to housing just because they will be less safe than other persons. This concept has been identified by behavioral researchers as the **right to risk concept**.

One of the most important findings from research on the public's acceptance of risk is that the public is much less accepting of risk when there is the potential of killing many people in a single incident. Other influences, such as whether those at risk have voluntarily assumed that risk, also affect its acceptability. Ultimately, this second factor is probably an important reason why fire safety requirements are lower for owner occupied residence than for equivalent rental units.

In the beginning of this textbook, we discuss the important relationship between roles and behavior. Here is another example. Roles profoundly affect how people view a situation. With regard to adult residential care facilities, fire prevention inspectors might not understand how owners have legitimate concerns about coursed containment and that this strongly influences their point of view. Simply, the facility owner feels confused and angry when presented with conflicting codes and cost estimates that differ greatly from fire company officers and fire prevention inspectors. Soon, the level of distrust is so great that these two groups find it impossible to work together. Ultimately, this is simply eliminated when fire company officers remain current with code revisions and remain on the same page as fire prevention inspectors and the buildings department.

Summary

An alarm signal may provide people with useful information about how they should respond, or it may add to their confusion and anxiety. For example, if a favorable fire alarm transmits an unintelligible message; this is likely to increase the anxiety of building occupants. Building performance during a fire also depends a great deal on "Fire Related Human Behavior." People use buildings in inappropriate ways that increase the risk of fire. Of significance, people cooking in hotel rooms have started fires. A change in occupancy in an existing building can create unforeseen problems when people use the building in ways the designers did not anticipate. Subsequently, storage of large amounts of combustible materials can overwhelm fire protection systems designed for ordinary hazards. We can divide systems models that usefully incorporate human behaviors into two basic categories, the hard systems approaches and soft systems approaches.

Among the earliest requirements incorporated into the building and fire safety codes were the design of exits and buildings. These requirements have evolved over time and rely on the judgments of code writers; therefore, some have criticized these codes for the lack of basis in research. Nonetheless, they have served good over many decades, despite their technical inadequacies. These codes are similar to hard systems approaches because they yield specific quantified results based on numerical input. However, it is worth noting that the system in question typically includes only a few of the components that determine how well people function during a fire or emergency. Also, they differ in that hard systems approaches incorporate clear rationales rather than unexplained judgments. The building design and its occupants should work toward a common fire safety goal. A quick and total evacuation is the logical goal in many buildings but what about high rise buildings, hospitals, prisons and the myriad of other occupancy types were a quick and total evacuation is not possible or desirable? In this chapter, you have learned about the concept of a life safety strategy as a goal towards which the performance of the buildings and occupants can work together as one system.

One of the most important findings from research on the public's acceptance of risk is that the public is much less accepting of risk when there is the potential of killing many people in a single incident. Other influences, such as whether those at risk have voluntarily assumed that risk, also affect its acceptability. Ultimately, this second factor is probably an important reason why fire safety requirements are lower for owner occupied residence than for equivalent rental units.

Discussion and Review

1. What are the two ways in which hard and soft systems approaches differ?

2. Hard systems models seem to provide objective results independent of the judgments of their users. One reason is that hard systems models can hide assumptions easily or otherwise make it easy for users to overlook assumptions. Can you give an example of such an assumption?

3. Can you list two ways in which hard systems models compensate for the difficulty of including human decision-making?

4. Does the Lane model given an accurate description as to how people really move down corridors and stairs?

5. Can you list, either, two reasons why the Lane model does or does not accurately reflect the way people really use stairs?

6. What are two factors that can increase the time needed to evacuate a building beyond the time that occupants need to simply travel along the egress routes?

7. The Lane model that was formerly used over estimates the carrying capacity of exits. Still, high-rise buildings have had a very low rate of fire casualties. Can you explain this discrepancy?

8. Do Fire Protection Engineers presently have a valid criterion for sizing their exit facilities?

9. To what does the term "design fire" referred to?

10. Can you name the computer model that is best described as a hazard assessment tool and is intended to represent the hazards associated with a single chosen fire scenario?

11. CRISP is a risk assessment model, what does this mean?

12. How does the CRISP model differ from other models that are deterministic?

13. What should be the common goal of the building design and its occupants?

14. Can you write two examples of life safety strategies?

15. Can you list two reasons for writing fire safety strategies?

16. Is it true that all future fire-related behavioral research will grow out on the experiences of the civilians at the World Trade Center on 9/11?

17. How does the model EXIT89 calculate walking speed?

18. True or false the definition of queued egress is when a number of people wait in line, or in queue, before entering the exit?

19. Can you describe the term known as hindsight bias?

20. Can you define the terms action schema and remote risk?

21. The term normative decision-making is also used in research literature and has the same meaning as what?

22. What is the scientific term for where something learned in one situation actually interferes with learning new procedures?

23. Can you explain the relevance to this course of NFPA code 101A, "Guide on Alternative Approaches to Life Safety?"

24. Can you explain the concept known as right to risk?

25. What really determines how quickly occupants can't evacuate a facility?

26. What is one of the reasons why people don't make better use of emergency exits?

4

Fire Safety Design and Fire Investigation

o o o o o o o o o o o o o o o o o o o o o o o o o o o o o o o o o o o o

"The study of uncontrolled fire appears to be motivated by clear risks to society and by society's having the means to invest in such studies."

—*Quintiere James G. in*
Principles of Fire Behavior

Introduction

In this chapter, we consider aspects of fire safety design and fire investigation. In fire safety design, a specific fire is not obvious; in fire investigation, we have some evidence or hypothesis of the fire. An important aspect of design is the specification of the relevant fire scenario and their probabilities of occurrence. In investigation, we may use our calculations to clearly eliminate or assert a certain event or its time of occurrence. I will discuss some examples and computer models that are often cited as the methodology for analysis.

Fire Safety Design and Local Regulations

Fire safety design is used principally in establishing compliance with the local codes. These regulations, which are based on codes and standards of practice, specify the requirements. The architect, contractor, builder or fire protection engineer must ensure compliance. In regards to the first three to do more is not usually sought; however, many situations arise, where the regulation is not clearly applicable or the designer wishes for an equivalent alternative. Subsequently, the establishment of **equivalency** requires at least a technical decision, but more appropriately an analysis. Ultimately, many aspects are subject to regulation and therefore open to design possibilities, including the following. The detection and alarm system, mitigation of growth and suppression, egress, continuity of operations, structural integrity, as well as refuge and rescue. I will not address all of these aspects in our discussion of the benefits of behavioral research to the fire service; because they are subjects of other textbooks.

Ultimately, the most significant aspect of fire safety design is the interaction of fire growth time with egress time. Moreover, we seek egress time to be less than fire damage time. Subsequently, for all possible fire scenarios and their probability or likelihood, we need to determine the egress time of the buildings population. Egress time increases as the fire progresses because smoke and fire slows down peoples escape. On the other hand, the time to cause damaged by fire is inversely related to the fire size. As the fire grows, the time for fire damage decreases. Of significance, sprinklers, suppression done by firefighters and fire resistant barriers all affect this damage time. It is imperative to note that sprinklers can be defeated and fire growth damage can take a turn for the worst again. The two examples of fire safety design analysis have actually been conducted because of fire incidents; they were not conducted as part of the design process. These examples are introduced to you the student to demonstrate that such an analysis is applicable in the design phase. The analysis of an actual fire incident demonstrates design strategies that could have been applied during development of the building, yet the building may have still been in compliance with the regulations after construction was completed. Subsequently, such analysis can provide cost-effective safety improvements or equivalent see to the regulations.

The following examples are cited as fire related behavioral research that was conducted by Professor James G. Quintiere. Today, there is a lot of talk about **performance fire safety codes** as opposed to prescriptive codes. Essentially this means using computational analysis over set specifications in the regulations. This subject was explained in previous chapters of this text and the student is reminded that this process is evolving and no fixed mythology has emerged. Moreover, because computations required a certain expertise, performance codes will not easily be put into practice until the knowledge is fully exchanged and appreciated at all levels of the process.

The first two examples of a fire safety design from regulation to computer were of the "Effects of Shaft Vents in a Building Fire" and "Smoke Movement in the World Trade Center, New York, New York."

The effect of shafts vents in a building fire was computed by Takeyoshi Tanaka, a scientist of the building research Institute of Japan, using his computer code design to calculate the smoke transport and its properties through a building. To shaft vent conditions are considered and lead to slightly different results for smoke movement and temperature. These computations were initiated following the MGM Fire of November 21, 1980, in Los Vegas, Nevada. At the time computers could not easily or quickly compute the 65 story MGM building. Subsequently, the calculations were only performed for a five-story building. Their differences were never used for analyzing that fire or learning from its consequences. Of significance, most of the 85 people that were killed in that fire were killed on the upper floors of the MGM as smoke traveled up the vertical shafts. The scientist Tanaka was able to show with his calculations that the case of a vent in the center shaft does not affect smoke temperature very much, but does ultimately impact the smoke depth on the third floor. Moreover, this computation implies further calculations could produce more complete information on the benefit of shaft vents.[30]

Another high-rise building fire attracted much attention, namely, the explosion and fire in the World Trade Center in New York City in April 1993. The computer code illustrated for this fire has since been updated and has been used to establish equivalent design alternatives in many Japanese building constructions. Computers now have much more capability and the entire 110 stories were included in the computations. These computations were implemented by behavioral scientists Yamaguchi and analyzed the fire scenario of the World Trade Center. A fire of 25 minutes was selected to simulate the parking garage fire. The results show the smoke flow at 13 minutes. No direct deaths resulted due to the smoke filling the towering building, probably because of the duration of the fire and the ventilation conditions. Such calculations could not only estimate the smoke conditions under this disastrous fire but could also be beneficial in assessing the hazard of other fire scenarios. Ultimately, design strategies that might make the building safer were evaluated on the computer.[31]

Until this expertise is comprehensive through the practice, fire safety design will continue to address equivalency issues. Changing fire safety design from regulations to computers is technologically feasible and in my opinion practical to implement. But, such new technology must progress hand-in-hand with needed research to fill the knowledge base and the computer gaps.

Group Response to Fire

Role behaviors are important during fires and emergencies. In general, people's roles tend to transfer effectively from routine to emergency situations. Competent people trained as part of the buildings emergency plan have considerable influence over how others perform. As a firefighter or someone involved in the fire service think about how your role influences your private life. For example, if you notice that a carbon monoxide detector is missing in your friend's house, are you likely to say something? Do your neighbors ever ask you fire prevention questions? If you are in a movie theater or in a place of public assembly and the fire alarm sounds, do you act any differently from the civilians around you? Can you think of another example of how everyday roles persevere during a fire or emergency? If you were acting as a fire consultant, what implications does this have in selecting a fire safety director or fire wardens and other members of an emergency response team who are supposed to give instructions? In the absence of trained personnel, such as a firefighter or fire safety director, other people often will emerge as leaders, perhaps as a function of their routine roles, knowledge, and personalities.

Fire prevention codes now commonly require verbal alarm systems in highrise buildings and with good reason. Verbal alarm systems provide explicit information to building occupants. People tend to respond much more quickly to verbal alarm signals. Explicit information overcomes the great shortcoming of the traditional fire alarm, which is inherently ambiguous. Nonverbal alarm signals provide very little information because they neither describe the exact nature of the threat nor provide instructions about how to proceed. Most people have learned over time that a fire alarm signal means that there is only a small chance that there is actually a fire even if they recognize the signal as a fire alarm in the first place. Subsequently, most people have heard many fire alarm signals during their lives, few of which indicated an actual fire they assume that there is a low probability that any single alarm signal indicates an actual emergency that could endanger their lives. Of significance, children typically respond better to alarm signals because they are used to following instructions without evaluating actual risk. Although verbal alarm systems generally a folk much quicker responses than nonverbal alarm signals, they must be used carefully.

In certain types of occupancies, finding an exit can be a problem. Fire Protection Specialists call this topic of trying to navigate in a building **"way finding."** As a firefighter or someone involved in the fire service, you probably have encountered this problem yourself. Especially in large, complex buildings, fire service personnel can have serious problems finding their way safely to attack fires and to rescue occupants. Often it is necessary to use floor plans, or someone very familiar with the building might have to direct you. However, the "way findings" method firefighters use may not work well for building occupants. For example, a study used people trying to travel to a room in a complicated office building, and compared various ways to help them find their way. The methods studied were a printed floor plans posted on the wall, verbal instructions from someone working in the building, directional signs posted on the walls of the building, and no help at all. Behavioral research found that the signs were the most effective. Verbal instructions were helpful, but much less so. Of significance, people actually performed worse with posted floor plans than with no help at all.[32]

The reason why signs are the most helpful method for finding your way and floor plans are the least helpful probably resulted from how much information people can store over a short period of time. In a complicated building, people have to memorize the direction of several turns, which is a difficult task. Moreover, the information is spatial; that is, occupants have to remember the relationships between objects in a physical space. They need to remember a sort of mental picture. This type of information can be difficult to process. Occupants did somewhat better with verbal instructions, probably because it is easier to remember verbal than spatial information. However, when following floor signs, people perform much better because they needed to remember only a single piece of information at any given time, a much more cognitively simple and straightforward task when faced with a fire or emergency.

As a firefighter or someone involved in the fire service or even maybe a fire victim or a once evacuated building occupant you are well acquainted with the effects of stress, both good and bad. A moderate amount of stress enhances performance. That rush of excitement increases your level of arousal and makes you more alert. You think faster and noticed more. But too much stress begins to degrade performance. It becomes more difficult to think clearly and to consider all the possible alternatives. It also becomes more difficult to attend to all of the information that may be relevant to the task at hand. Think about your own experiences and try to remember when the level of stress was so high that it interfered with your ability to notice clues not related directly to your immediate task. This might have occurred during a firefighting operation or perhaps you narrowly avoided an automobile accident while arguing with someone as you cross the street.

Every time the fire alarm signal activates and there is no real threat, the likelihood that people respond to a future alarm is reduced. This is called a **false alarm affects.** False alarms and malicious alarms as well as surprise fire drills all can contribute to the problem, because each type of event tends to weaken the perceived relationship between an alarms signal and an actual threat. The most obvious solution is to reduce the number of false and nuisance alarms. New technology for smoke detectors can reduce the number of false alarms by making it possible to better differentiate nuisance alarms from real fires. Next, the responsible party should cancel nuisance and false alarms as soon as possible. In the case of a malicious alarm, building managers should allow occupants to return as soon as possible but not until the fire department verifies that there is no significant threat. Third, the system should employ differentiated warning signals to indicate the degree of a potential threat. For example, in a high-rise office building, an announcement can inform occupants that the management has received an alarm and that occupants should stay on the alert in case an evacuation is ordered by the fire department. Of course, if it turns out that there is no real threat, building officials must inform occupants as soon as practical. Fourth, people need to know the different courses of fire alarms. For example, occupants should know whether a system malfunctioned coursing the false alarm, or whether smoke from an overheated electrical motor or a minor fire caused a fire alarms. This will help people understand that there are several potential causes of alarms and any single alarm could indicate a serious fire. Lastly, building occupants need to know how those in charge or correcting the source of a nuisance or malicious alarm. This will reduce the perceived probability that the same problem may cause another false alarm.

Fire Science Behavioral Researchers have studied power and influence extensively. In this situation, the fire department has influence and power over the building managers but neither the fire department nor the building managers has a great deal of power over the occupants. It is easy to exaggerate the amount of influence that building managers have over the behaviors of their tenants. In theory, they may be responsible for seeing that tenants participate in fire prevention programs. In practice, they are not directly responsible for enforcing fire prevention regulations and are unlikely to motivate tenants by turning them in to fire department prevention officers. However, building managers can strongly influence the behaviors of their tenants. When tenants perceive management as concerned and helpful in reducing the risk of fire loss while remaining sensitive to their needs, they are likely to be cooperative and at times grateful.

Most people commonly believe that the smell of smoke will wake them up in the event of an actual fire. However, recent evidence indicates that this is not true. In fact, people may be less likely to wake up to smells than to sounds. A recent study examined the likelihood that the smell of smoke will wake people. Researchers subjected 10 subjects in a laboratory to odors, including smoke but only two of them woke up. While the sample size and controls in this study are not definitive, the findings offer support for the hypotheses that the smell of smoke, by itself, is unlikely to awaken people.

Building occupants often frustrate fire safety professionals when they fail to leave upon hearing an alarm. However, fire safety professionals are wrong to attribute such behavior to stupidity. People typically delay their responses to ambiguous signs of danger until they understand the threat better. Fire Science Behavioral Researchers suggest that good training and accurate information early in the fire incident would alleviate much of this problem.

In buildings where the actual occupancy loads approach the design occupancy loads, crowding will occur during emergency egress when everyone attempts to leave at the same time. Behavioral researchers studying fires can use different approaches to reduce the likelihood of crowding at exits. Behavioral scientists have conducted considerable research on ways to compute the carrying capacity of exits. As you have seen in the previous chapters of this book methods are available for calculating flow rates through crowded paths of egress. There are two basic approaches: a formula derived from empirical studies of actual flow rates in office buildings; and calculation methods based on physical models that draw an analogy between hydraulic flow rates, liquids flowing through pipes, and the flow of people through crowded paths of egress. The former method rests on studies performed under the direction of Jake Pauls for the National Research Council of Canada and is available in the National Fire Protection Association's Guide on Alternative Approaches to Life Safety Code as an alternative method. The second approach is the basis for some computer models that fire protection engineers used to estimate best evacuation times. Of significance, if people more commonly adapt and use performance-based codes behavioral researchers will use such calculation-based methods with more frequency. However, these approaches have limited usefulness because they measure the best evacuation times that occupants are likely to achieve. The methods do not measure the time that building occupants take before starting to evacuate or inefficiencies in managing the evacuation. Previously in this book's chapters you read about the use of hard and soft systems methodologies that largely involved how best to estimate evacuation times from buildings. It remains that two import methods for reducing crowding or to phase an evacuation and to relocate occupants temporarily within the building.

There is good evidence that people tend to congregate in groups during fires in large buildings. University of Maryland first reported these convergence clusters in an investigation of human behavior during a fire in a high-rise apartment building. Further research of the disastrous fires at the MGM Grand Hotel in Las

Vegas, in which 85 people died and West chase Hilton Hotel fire in Huston, in which 12 people died, confirmed that people tend to gather in these convergence clusters during large fires. Think about why people tend to congregate into groups during fires. Ultimately, people indicated that being with other people helped to reduce stress, and they also chose areas that offered access to balconies and had a relatively better conditions. You might think about how firefighters can use convergence clusters to their advantage during a fire in a large apartment building or hotel. For example, how could you plan a more efficient rescue of people trapped inside a high-rise office building now that you know about convergence clusters?[33]

Currently, modern buildings are designed to provide good access for people with disabilities. Elevators allow them to reach nearly every occupied location. But getting out during a fire is another problem entirely, especially when using stairs is impossible or very difficult. In most fire incidents, many disabled people in a building experience lengthy delays before they are evacuated. Think about people with disabilities, people in wheelchairs immediately come to mind. However, of the people who have problems during emergency evacuations only a small portion are wheelchair users.

Firefighter Fatalities Reports

The National Fire Protection Association (NFPA) is responsible for many articles on firefighter deaths over the years, but in 1977 a concentrated effort was made to identify all on-duty fatalities that had occurred in the previous year. Since then, NFPA has conducted an annual comprehensive study of on-duty deaths in the United States. Recently, the NFPA marked the 30th year of this study and this is why I find it significant enough to provide it in this book as an opportunity for you to see how things have, and maybe, have not, changed over 30 years.

The average number of annual on-duty firefighter deaths has dropped by one third over the past 30 years. In the late 1970s an average of 151 firefighters were killed on duty annually. By the 1990s, the average had dropped to 97 deaths per year. So far in this first decade of the 21st century, the annual average has held steady at 99 deaths per year. There have been four years with firefighter death tolls below 90. In 2006, 89 firefighters died as a result of on-duty injuries. There are two major forces driving these decreases in death. First is a drop in the number of on-duty fatalities attributed to sudden cardiac death. Deaths due to or from apparatus, which claimed at least three deaths in most of the first 11 years, virtually disappeared in the 1990s. One such death has occurred in three of the past four years, however. I will discuss these and other trends in more detail in this brief review of the past 30 years that the NFPA has been studying Fire Related Human Behavior.

The first five years of the National Fire Protection Association's study, an average of 65 on duty fatalities each year was due to sudden cardiac death. In the most recent five years, the annual average dropped to 41. Sudden cardiac death is defined by the American Heart Association on their website www. americanheart.org as "the sudden, abrupt loss of a heart function in a person who may or may not have diagnosed heart disease." The number of sudden cardiac deaths annually has fallen by approximately one third from the late 1970s; however, since the early 1990s, the number of deaths each year has tended to fluctuate between 40 and 50, with no clear trend up or down. There were 34 sudden cardiac deaths in 2006, the lowest number recorded over the 30 years of the NFPA study. The largest proportion of the victim's experienced cardiac symptoms during fired ground operations. The largest proportion involved firefighters responding to or returning from alarms. In its investigation of on-duty cardiac related fatalities, the National Institute of Safety and Health or NIOSH reports, "firefighting activities are strenuous and often require firefighters to work at near maximum heart rates for long periods. The increase in heart rate has been shown to begin with responding to the initial alarm and to persist through the course of fire suppression activities."[34]

Since 1977, the number of US deaths annually at structural fires has dropped 69%. In the late 1970s, an average of approximately 60 deaths occurred each year at structural fires. That number dropped to an average of 20 per year in the past three years. This finding has often been credited to improvements in protective clothing and equipment, fire ground command and control procedures, and training; and although those changes certainly played a role, little attention has been paid to the drop in the number of structural fires themselves. While deaths at structural fires have dropped 69%, the annual number of structural fires declined by 53%. **To what degree, then, has the decrease in firefighter deaths been driven by the drop in the number of structural fires?** This comparison of the decline in the number of structural fires in the decrease in the number of firefighter deaths at structural fires shows that the trends track fairly closely, indicating that the drop in deaths may have been, to a great degree, as a result of the reduction in the number of fires. **This leads to an important second question; are firefighters just as likely to die at structural fires today as they were 25 or 30 years ago?** The one area that had shown marked increases during the period is the rate of deaths due to traumatic injuries while operating inside structures. In the late 1970s, dramatic deaths inside structures occurred at a rate of 1.8 deaths per hundred thousand structural fires and by the late 1990s had risen to approximately 3.0 deaths per hundred thousand structural fires. Since that time, the rate has fallen and now stands at 1.9 deaths per hundred thousand structural fires, a rate only slightly lower than that observed in the early 1980s. Almost all of these non-cardiac fatalities inside structure fires were the result of smoke inhalation, burns and crushing or internal trauma. A detailed look at each incident is beyond the scope of this book, but the National Institute for Occupational Safety and Health has a program of on-site data collection and investigation of on-duty fire-fighter fatalities that provides a valuable database. Reports on many of the most recent fatalities can be found on their website: www.cdc. gov/niosh/firehome.html.

The largest share of fire ground deaths in 2006 occurred during forestry fires. This is only the second time in the 30-year period that wildland fires accounted for the largest share of fire ground deaths. Just as the severity of wildland fire season varies from year to year, the number of firefighters dying in wildland fires also varies. Of significance, wildland fires frequently claimed a large number of fire-fighter's lives in a single incident and almost half of the victims of wildland fires were volunteer firefighters, followed by contract is for state and federal land management agencies, employees of federal land management agencies, and employees of state land management agencies. The remaining victims were career firefighters, members or supervisors of a prison inmate crew, or military or industrial firefighters.

Deaths in road vehicle crashes over the past 30 years consistently account for the second largest share of firefighter deaths, overall. These crashes occurred during all types of on-duty assignments, not just while responding to or returning from alarms. Three quarters of the victims in these crashes were volunteer fire-fighters. Fourteen percent were career firefighters and the remaining victims were contractors for, or employees of, state and federal land management agencies. Of significance, more than one third of the deaths involved firefighter's personal vehicles. Obeying traffic laws, using seat belts, driving sober and controlling driving speeds would prevent most of the firefighter fatalities in road crashes each year. Two NFPA standards are available to help fire departments established safe driving programs: NFPA 1002, Fire Apparatus Driver/Operate a Professional Qualifications, and NFPA 1451 Fire Service Vehicle Operations Training Program. NFPA 1002 identifies the minimum job performance requirements for firefighters who drive and operate fire apparatus, in both emergency and non-emergency situations. NFPA 1451 provides for the development of a written vehicle operations training program, including the organizational procedures for training, vehicle maintenance and identifying equipment deficiencies. In addition, NFPA 1911, Inspections, Testing, Maintenance, and Retirement of In-service Automotive Fire Apparatus, details a program to ensure that fire apparatus are serviced and maintained to keep them in safe operating condition.

From 1977 to 1987, a total of 41 career or volunteer firefighters died, at least three each year, when they fell from apparatus while responding to or returning from alarms. The victim may have been standing, donning his gear, when he fell from the apparatus. Investigators believe an unlocked crew door is to blame in most instances. **Moreover, an enclosed crew cab riding area does not guarantee protection.** Ultimately, fire departments should ensure that all interior crew and driving compartment door handles are designed and installed to protect against inadvertent openings and departments should consistently enforce the requirement that all firefighters responding and returning from alarms should be restrained by seatbelts or safety restraints at all times the vehicle is in motion.

Deaths during training activities account for 7.4% of all on-duty firefighter fatalities over the past 30 years. Moreover, the disturbing fact is that firefighter deaths during training are particularly needless, as the purpose of training is to prevent deaths and injuries and should certainly not be the cause of fatalities. The larger shares of training related deaths occurred while the victims were participating in apparatus and equipment drills and while firefighters were taking part in physical fitness training. Just over half of the firefighters who died while training died due to cardiac events. Dramatic injuries, smoke inhalation and drowning what the next three major courses of training deaths.

Fire Analysis for Fire Investigation

Litigation has pushed the use of computers and fire analysis into the field of fire investigation. As in design applications, there is no simple approach without a competent knowledge of fire related behavior. Although computer models can be useful, the knowledge gained in this textbook can go a long way to bolster a time-line or proposed scenario in a fire investigation. Unlike design, this process for investigation is not open ended. Because, there was a specific fire and it did specific damage. Ultimately, these facts must be established through evidence, witness accounts and analysis. I will examine some specific cases to give you specific examples that illustrate the process of fire analysis for fire investigation.

On August 3, 1978, a fire occurred at a Waldbaum Supermarket in Brooklyn, New York. At the time of the fire, an extension was under construction. The store was a typical supermarket with a mezzanine along a portion of the north wall. The loft was completely of wood construction compromising the floor, roof and structural supporting trusses. The trusses were made of 3 x 12 inch members interlaced together in bundles of four or five. Some trusses were covered on one side with plaster to form firewalls in the loft. However, to enable passage, these trusses had doorway openings. The roof had been modified within an added rain roof at the peak, which formed a double layer of roof at that peak. Also the new construction required a splayed roof section to meet the new roof of the extension, forming another double roof triangular section along the north wall.

Routinely, construction work began at 7 a.m. and the store opened for business at 8 a.m. Fire was first reported at 8:30 a.m. along the interface between the ceiling and extension wall of the mezzanine men's room in the machine room. The fire eventually spread into the loft between trusses and resulted in a trust collapse in the loft at approximately 9 15 a.m., causing 12 New York City firefighters to fall into the flames, ultimately killing six of them. Of significance, a man was tried and convicted in 1978 of setting this fire. His confession stated that he and two others set the fire near dawn at approximately 6 a.m. by making holes in the roof and using newspaper and lighter fluid to initiate the fire below. This confession was later questioned and discounted in a retrial that was held in 1994. A consistent fire scenario was never fully presented. The original fire marshals could not agree on a cause and later suggested that the cause of the fire may have been of electrical origin. However, there was no electrical power in the ceiling area where flames were first seen and a man in the loft at the onset of the flames saw no evidence of fire. Ultimately, the District Attorney in Brooklyn sought advice on how to explain an alleged fire starting at 6 a.m. but not being seen until 8:30 a.m.

William J. Petraitis, a special Agent of the Bureau of Alcohol, Tobacco, and Firearms reinvestigated this fire 16 years after it occurred. He discovered the slated roof extension and reason the fire had to begin in that space. Then he examined the hypotheses of the fire beginning there at approximately 6 a.m. to see if it could be consistent with the other known events. A fire scenario was developed in calculations were examined to support the plausibility of the events and their duration. I will not present all of the analysis in this textbook but I will describe the results.

At approximately 6 a.m., and arsonists is considered to have set a fire in the roof area adjacent to the construction of the building extension. The modus operandi is stuffing paper through holes in the new roof extension along with a liquid accelerant. The Splayed roof extension has been built over the existing roof and forms a void space between the two roofs. The new roof is supported by rafters. The fire is set in channels of the wood rafters that extend between the primary wood trusses. Gaps under the rafters allow air to flow into the fire but the space is mainly enclosed with temporary partitions at the wall adjacent to the new building extension. It is estimated that the accelerant soaked paper caused a fire of a 100 to 500 kW in one or more rafter channels, involving no more than 1 m^2 of fire area. Under expected heat fluxes of 40 to 50 kW per square meter, the wood members would ignite in 30 seconds to one minute and begin to continue burning at roughly 500 kW. The accelerant fire would probably burn out in one to two minutes. The wood fire could progress to roughly 1000 kW but then it would become limited in further growth due to combustion products filling this confined space. Oxygen depletion would cause flaming to cease in about one to two minutes following wood ignition. However, sustained smoldering would remain especially in the small rafter spaces where radiant heat transfer would be high. Some time around 8:30 a.m. a hole would be cut between the loft and the roof cavity. Ultimately, with air velocity increased the smoldering wood change to flaming combustion. Such velocities would be realized at the hole as air was naturally introduced to flow from the loft into the hole. Flames would erupt in the false roof cavity space. Subsequently, at this time flames are observed in the mezzanine area at the wall and ceiling near the trusses. It is believed that this was due to the expansion of the flames as flashover caused a rapid rise in the energy releasing rate in this confined space. Moreover, the associated pressure increase forced flames through available void spaces. As the hole to the loft became larger, the increased airflow rate would also promote a larger flaming fire in the rafter spaces. The flow of the flaming combustion products would flow the rafter channels to the endpoints of each of the trusses. Moreover, as the hole to the loft became even larger, flames would lap upward under the sloping wood ceiling of the loft. The person in the loft before this time would not have been aware of the fire because smoke would not permeate into the loft until the hole became large enough. In fact, this witness only saw flames when he descended to the machine room after being warned of the fire.[35]

Flames spread would rapidly move from the hole to up and under the loft wood ceiling. It is estimated that flames would move from the hole region up and along the peak of the loft to the trusses within two to three minutes. This rapid spread seemed to credible for the investigator so an experiment was performed in a similar, but small loft. Flashover occurred in the independent experiment in approximately 1 minute. The flames spread calculations that were acquired in the independent study were then apply to this fatal fire study and indicated that flames spread would occur in approximately 2 1/2 minutes. From the onset of flaming at the hole, full involvement of the loft section would take only five to ten minutes. Moreover, because of the false roof sections in the trust firewall sections, the firefighters on the roof were unaware of the raging fire in the loft section below them.

This fire scenario was a complex series of phenomenon that are not usually appreciated in the field of fire related behavior. Experienced fire marshals were not able to deduce this process. Subsequently though a relatively simple scientific analysis was able to produce a plausible and consistent series of events within the periods of the recorded observations on this fire. Ultimately, this scenario can not be described or defended without a comprehensive knowledge of fire related behavior and available data to support the calculations. The flashover and smoldering research data are rare but are critically needed in fire behavior analysis. Moreover, this analysis has chocked open the door to future fire investigations.

Behavioral Factors of Arson

When investigating a suspicious fire an investigator relies on the presence of three elements. They are burned property, intentional burning, and malicious ignition; that is, with the intent to harm or destroy. People start many fires with good intentions they include fires to cook, provide heat or for advanced work like plumbing. Generally, investigations of these fires conclude with the **human era** and not **malicious intention**. When an individual or group starts a fire intending to damage or destroy property, the perpetrator or perpetrators' has in fact committed arson. The National Center for the Analysis of Violent Crime (NCAVC) studied the inner drives that cause a person to commit arson. The NCAVC has classified the motives of arson fires into six major categories. They are vandalism, excitement, revenge, crime concealment, profit and religious extremists. Fire behavioral researchers think that factors outside of this framework motivate juveniles to set fires. Experimental fascination with fire is a very normal and routine experience among children. As was explained previously in this textbook children can become curious about fire, igniting a blaze that unintentionally grows beyond control. Moreover, this is often the case with very young juveniles even those as young as two years old. Additionally, a level of juvenile fire activity involves children who are angry or seek attention. In effect, these fires represent these children's call for help. Ultimately, their anger or need for attention may result from domestic upheaval, such as the death of a family member or other loved one, divorce, loss of a job, substance abuse, relocation, or even incarceration. Subsequently, it can also develop from abuse and or sexual or psychological. The **"Paris Hilton" syndrome** which includes absent or inattentive parents, including those considered successful, also can create an atmosphere that causes juveniles to use fire as a means for seeking attention. Moreover, social problems or difficulties in school also can be the basis for acts of this nature.

A very small percentage of the juvenile fire setter population exhibits persistent, aggressive antisocial behavior. These children demonstrate a history of unsupervised fire starts, the primary purpose of which is miscellaneous mischief. The fire often targets the property of others, and the child will not admit responsibility for the fire readily. Fire attracts these types of juveniles and they try to watch the fire burn.

Fire marshals usually classified fires set at random and with no identifiable purpose as **vandalism related**. The arsonists typically are juveniles or adolescence target properties that are normally within their environment. Examples of such targets include but are not limited to schools and adjacent buildings, abandoned or vacant buildings like garages and sheds, older vehicles or vehicles left parked for long periods, brush and other vegetation wild-lands. The arsonists may act alone or in conjunction with others in the group. Ultimately, it is time proven through behavioral research that boredom and frustration on known triggers in juvenile arson cases. Although these fires can seem totally unpredictable and pattern-less modules have success by tracking vandalism fires by place of occurrence and time of occurrence.

Research shows that the need for attention drives the **excitement fire setter**. The attention can come from the excitement of the incident, recognition, and very rarely, sexual stimulation. During the fire, the arsonists most likely will also stay in the area to watch, this is why excitement fire setters usually set fires close to their home. The arsonists may also engage in participatory activity such as photographing or videotaping the fire. A subset of the excitement fire setter is the **vanity fire setter**. These fires are ignited to provide the individual with an opportunity to take action and be noticed. One of the most concealed facts regarding the "Son of Sam" case was the fact that David Berkowitz was a New York City Auxiliary Police Officer. During his voluntary patrol duty the "Son of Sam" AKA David Berkowitz would routinely come across private dwelling house fires in specific geographical areas in New York City's borough of Brooklyn. The modus operandi for these and numerous dwelling fires was a fire that was set under the front porch immediately after the family occupancies gathered for dinner. During the incipient stage of fire or shortly thereafter depending on if the occupancies noticed the fire themselves or not, David Berkowitz would show up and call in the fire over his provided police radio, ultimately taking credit for alerting the occupants. Prior to his arrest as the "Son of Sam" killer New York City Fire Marshals were actively investigating him and were close to arresting him before their case was shut down and taken over by the New York City Police Department.

The motive responsible for perhaps the greatest number of fires is **revenge**. A person, acting alone ignites a fire in retaliation for an insult, real or imagined, or perhaps in reaction to some form of hostile behavior he or she has experienced. Ultimately, this type of arsonists believes someone has wronged him or her in some way and sets a fire to get even! New York City Fire Marshal statistics show revenge fires occur as a result of scorn lovers, marital infidelity, hostile relationships and racial discord. Subsequently, landlord/tenant disputes can precipitate these fires as well.

There is a common misconception that in a fire, especially a large fire, the conflagration destroys physical evidence. Although the fire may damage the evidence and even destroy some, it will not destroy all the physical evidence. The arsonists motivation and the type of arson he or she commits will yield some insight into what he or she burns and why. For example if the crime is of domestic violence that results in a person's death, the arsonist's has probably set the fire to cover the attack on their domestic partner and the suspect pool is narrowed.

Arson for profit is defined as a fire set to create an unlawful gain, either directly or indirectly. The number of potential scenarios is as endless as the creativity of those who would use arson to defraud, extort and even intimidate. Some reasons why people set arson for profit fires is to collect insurance, eliminate business competition, extort money, improve property value, eliminate obstacles to development, gain higher standing on welfare rolls, get out of an unprofitable lease or contract, or to create a need for services as a result of the fire.

The most commonly utilized weapon by terrorists throughout the ages as well as presently is fire. **Extremist set fires to terrorize**, to instill fear and to impose their beliefs on others. That fact has not changed, even in modern times when anthrax is a common kitchen table discussion. Examples in recent times include fire bombings of abortion clinics and houses of worship. In many cases, these arsonists target a structure in order to maximize the impact of their message. If the message is antigovernment, then the target might be a large federal building. An extension of this may be to suggest vulnerability and so the target may be a federal building in a relatively peaceful area, like we saw in Oklahoma City. Currently in Iraq and Afghanistan arson is the number one tool being used by insurgents to attack our troops. The evil insurgents like to use improvised explosive devices that give of large flashes and result in sustaining combustion because fire and smoke draws the media and the press enables the fear factor. Only the beliefs of the arsonists limit the targets they choose.

Summary

Computer models are mathematical solutions implemented and displayed on computers. They attempt to alleviate the user from the direct development of the mathematical solution and usually make it convenient and easy for the person using the program. Of significance, if the user wishes to solve a problem outside the scope of the original model this can lead to serious misinterpretations. Ultimately, the use of computer models requires an understanding of fire dynamics and knowledge of the model. It is the author's recommendation that a computer model should only be a convenient tool, not a user-friendly device for any use. The fire protection specialists using fire computations must begin with an understanding of the process associated with the fire. Then the mathematical tool must be understood. Moreover, only then can the use of computer models truly benefit the fire service.

Discussion and Review

1. What is the overt difference between a fire safety design and a fire investigation?

2. What is the important aspect of design?

3. In an investigation what do we use to eliminate or assert a certain event or its time of occurrence?

4. True or false, there is much talk about shifting from performance fire safety codes to prescriptive codes?

5. Is the performance fire safety code process of evolving?

6. Do you agree or disagree that because fire computations required certain expertise, performance codes will not easily be put into practice until the knowledge is fully exchanged and appreciated at all levels of the process?

7. Can you name one regulation in fire safety that could be enhanced by analysis?

8. What type of research is critically needed in fire behavior analysis?

9. When investigating a suspicious fire an investigator relies on the presence of what three elements?

10. The NCAVC has classified the motives of arson fires into six major categories, what are they?

11. What is the "Paris Hilton" syndrome?

12. True or false, a very small percentage of the juvenile fire setter population exhibits persistent, aggressive antisocial behavior?

13. Fire marshals usually classified fires set at random and with no identifiable purpose as vandalism related, is this statement true?

14. What is a subset of the excitement fire setter?

15. What is the motive responsible for perhaps the greatest number of fires?

16. True of false in general, people's roles tend to transfer effectively from routine to emergency?

17. Is there any good evidence that people tend to congregate in groups during fires in large buildings?

18. In certain types of occupancies, finding an exit can be a problem. What do Fire Protection Specialists call this topic of trying to navigate in a building?

19. Can you explain false alarm affects?

20. Is it true that most people commonly believe that the smell of smoke will wake them up in the event of an actual fire?

21. What do Fire Science Behavioral Researchers suggest as a way to alleviate a typical delayed response by occupants to ambiguous signs of danger until they understand the threat better?

APPENDIX A

Short Staffing: A Congressional Priority

It's a disgraceful fact that for the past 30 years the staffing of fire department apparatus has continued to decline. There are presently is a countrywide practice of sending two or even one firefighter on an apparatus to an emergency call! To address this issue, which is the most significant problem facing America's firefighters, it is my opinion that we have to focus our political energies locally. Moreover, for highly effective firefighting to be orchestrated the right combination of properly placed apparatus staffed by well drilled physically fit firefighters led by experienced and responsible officers must be assembled. These firefighters need to be assembled quickly with a preconceived standard operating procedure in place. Without a sufficient number of firefighters, you can't adequately address stress levels, task allocation, tactical integration and FAST teams, or anything else that might be needed to prevail over our historically consistent enemy of fire. Of significance, the real issue in fire ground safety begins with staffing and not technology.

The National Fire Protection Association standards 1710 and 1720, the organization and deployment standards for Korea and volunteer fire departments, respectively, represent the only acceptable staffing levels for structural firefighting. I fully understand that in many places it is not now possible to meet these standards. This unfortunate reality sometimes is based on rural community's lifestyles but often it simply is a reflection of the acceptance of substandard fire safety by elected officials. Nevertheless, the standards are not law unless your community chooses to adopt them as such. The important thing to remember here is the NFPA standard staffing number was not based on a commercial fire; it was based on a 2,000 square foot residential structure. This simple structure represents the most common fire-ground in America. NFPA 1710 and 1720 were as controversial when they were presented as NFPA 1500 (Standard on Fire Department

Health and Wellness) was before them. Unfortunately, mayors, city council and city managers knew when 1710 and 1720 were passed that they would ignore them. This indifference is based on a faulty assumption that they will never be faced with a significant fire. They will support hose testing, plump testing, and anything else that is definable and has a short-term dollar amounts, but staffing or health and wellness are non-entities. We must increase our local pressure on these policy makers. We can show them task by task how labor intensive the fire ground is and then prove to them that at least seven of these tasks need to be happening simultaneously to make a positive impact on "Fire Related Human Behavior."

Notably, the staffing crisis was escalated federally when the Bush administration proposed zero-funding for the SAFER Act in 2008. The SAFER Act was enacted in 2005 to provide funding for the hiring of 75,000 new firefighters nationwide, and is a kind of personnel buy-down plan. SAFER, was designed to help departments hire firefighters by supporting a portion of their salary over five years. This gives a department the firefighters it needs as well as time to develop a budget to support them long term. Although the financial burden for fire protection is a local issue, this plan makes it possible to ease the resistance to securing staffing standards. The current federal budget proposal reflects on the national level the viewpoint on firefighter staffing we all find so intolerable on the local level. Aside from complete elimination of SAFER, the 2008 allocation of the Fire Act is almost $400 million less than 2005; Rural Fire Assistance Program is gone completely; and the Hometown Heroes Act, which was passed in 2003, has yet to pay a single dime in benefits. Hometown Heroes is a very important recruiting tool for volunteer firefighters and is all about achieving safe staffing.

We can turn this situation around by putting pressure on congressional representatives to get our funding restored. To do this, they have to hear from us. We need to use every tool available to explain why every member on every piece of equipment is critical to the tasks a company performs. There is evidence that many cardiac cases are directly related to overexertion on-the-fire ground because of substandard staffing. The ability to ladder, ventilate, search, extinguish a fire and provide EMS services when needed is hamstrung by substandard staffing. Moreover, every single life saving, time critical activity we can identify is adversely affected by substandard staffing.

It is my personal opinion that we must do a better job of explaining the critical coordination aspects of our tactical activities. Our elected officials think linearly, in sequence, so you have to show that on the totally integrated via ground 1, 2 and 3 are all happening at the same time. The prize is getting your local offi-